Ania Wyczalkowski ▪ Ro‍
George Mason Univ‍

LAB MANUAL

to accompany

Physics
Matters

James Trefil
Robert M. Hazen
George Mason University

WILEY

JOHN WILEY & SONS, INC.

Contents

Name: Nathan Smith   Date: 10/22/07

Forces in 2–Dimensions

Purpose

- To further investigate the concept of force, Newton's Law's, and to practice vector addition in two dimensions.

Equipment and Supplies

- ☐ Spring scales (4)
- ☐ String
- ☐ Metal ring (e.g. key chain ring)
- ☐ Scotch tape
- ☐ Ruler

INTRODUCTION

In a previous lab you learned about forces and their addition in one dimension. Here we will consider a two dimensional case where the forces do not act along one particular line.

Imagine that you have an object that is acted upon by two forces $F_1 = 5N$ and $F_2 = 3N$ which make an angle $\alpha = 60°$ to each other as shown in figure 5-1 below. Here each vector is represented by an arrow which length represents this force's magnitude.

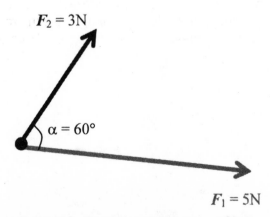

Figure 5-1. Two forces acting on an object.

To find the net force we need to add these vectors. To do that we copy vector F_1 and F_2 as shown in figure 5-2a. While copying we must retain the original direction and length of each vector. You can think of this copying procedure as sliding an arrow. The resulting copies shown dashed should be parallel to the original vectors.

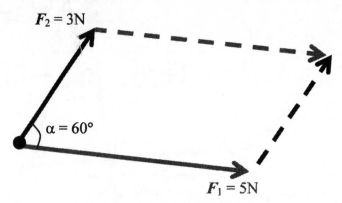

Figure 5-2a. Adding two vectors.

Now you can see that the original vectors and their copies form a parallelogram. The net force $F_{net\ 12}$ vector is represented by a diagonal in this parallelogram as shown in figure 5-2b.

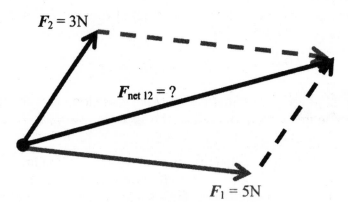

Figure 5-2b. Adding two vectors.

You can use the ruler to convince yourself that the length of the diagonal in our example is about 14 cm and thus the magnitude of $F_{net\ 12}$ = 7 N, given the arbitrary scale we are using where 1 N = 2 cm.

Question

a) In the space provided below in figure 5-3 add vectors **A** and **B**. Draw the resultant vector and find magnitudes of **A**, **B** and the resultant vector if the scale is 1 cm ~1N.

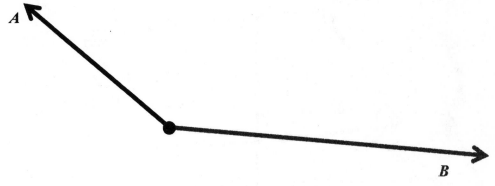

Figure 5-3. Two vectors.

PROCEDURE

PART 1 ■

1. Tear out the sheet of paper with figure 5-4 out of your manual and tape it to the middle of your lab table.

2. Tie three pieces of string to a metal ring. Make loose loops around the ring so that the strings can slide freely around it. Make the loops long enough so that they extend to or beyond the edges of the sheet with figure 5-4 when you stretched them along the three thick lines drawn in the figure. These lines make 120° angles with each other.

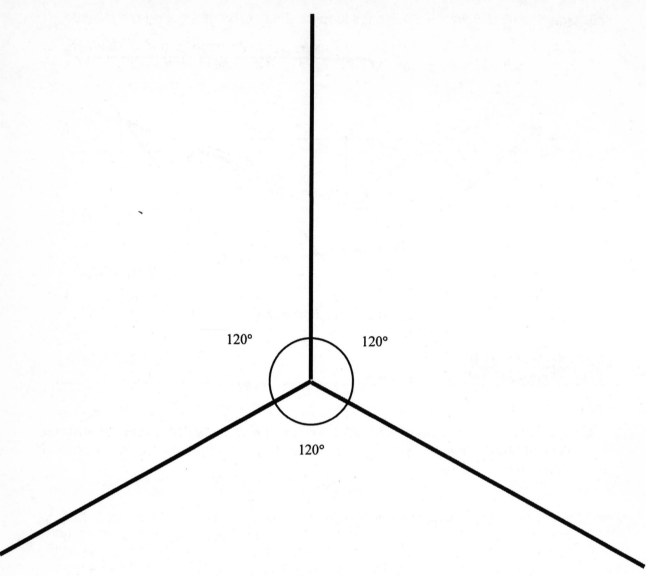

Figure 5-4.

3. Attach the other ends of the string's loops to three spring scales as shown in figure 5-5.

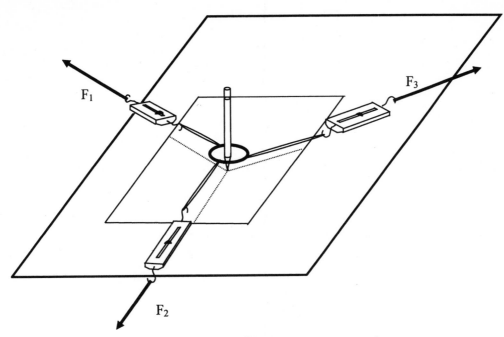

Figure 5-5. Apparatus for PART 1.

4. You and your partner now pull on the scales. Keep the scales horizontal. Can you keep the ring stationary if all three scales pull to one side rather than at 120° to each other? Why? Discuss.

5. Can you keep the ring stationary if two scales pull pretty much to one side and one in opposite direction? Why? Discuss.

6. Spread the scales in three different directions (making 120° angles) along the thick lines drawn in the figure 5-4 and pull. It is important to align the center of the metal ring with the intersection of the lines as well as to align the strings with the thick lines. Also make sure the whole apparatus it held horizontally over the table.

7. Record the readings of the three spring scales. Make sure not to confuse which scale reading goes with each scale.

$$F_1 =$$

$$F_2 =$$

$$F_3 =$$

8. Do you notice anything particular about the magnitudes of these forces?

9. In figure 5-4 draw the horizontal forces acting on the ring. Assign a scale suitable to make a large clear drawing yet not run out of room (e.g. 1cm ~1N or 0.5 cm = 1N). Label the vectors.

10. Find the resultant force from adding F_1 and F_2

$$F_{net\ 12} =$$

How does its magnitude compare with F_3?

How does its direction compare with F_3?

11. Do $F_{net\ 12}$ and F_3 balance each other?

12. Given the accuracy of your measurements have you verified Newton's 1st Law? Does the result meet your expectations? Discuss.

PART 2 ■

1. Repeat procedure outlined in **PART 1** (without repeating **PART 1.4** and **5**) but now use sheet in figure 5-6. The angles between the thick lines are indicated on the sheet.

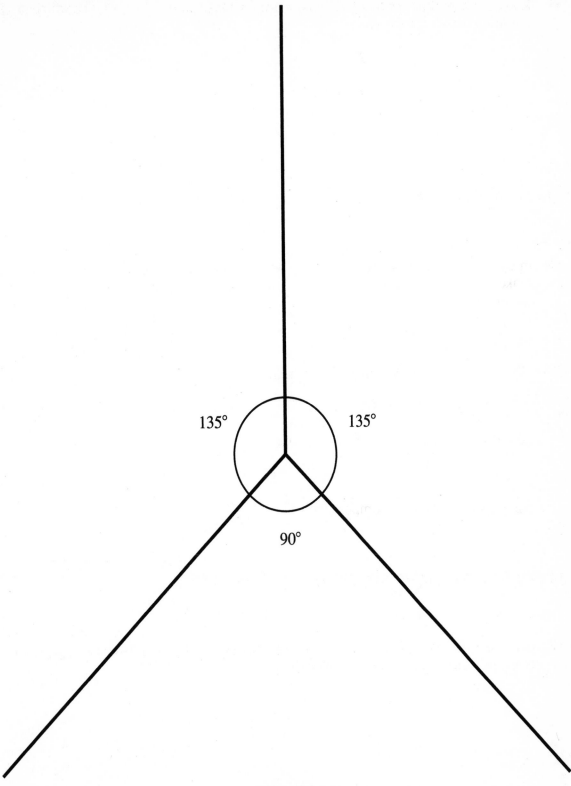

Figure 5-6.

2. Record the readings of the three spring scales. Make sure not to confuse which scale reading goes with each scale.

$$F_1 =$$

$$F_2 =$$

$$F_3 =$$

3. Do you notice anything particular about the magnitudes of these forces?

4. In figure 5-6 draw the horizontal forces acting on the ring. Assign a scale suitable to make a large clear drawing yet not run out of room (e.g. 1cm $\sim$1N or 0.5 cm = 1N). Label the vectors.

5. Find the resultant force from adding F_1 and F_2

$$F_{net\ 12} =$$

How does its magnitude compare with F_3?

How does its direction compare with F_3?

6. Do $F_{net\ 12}$ and F_3 balance each other?

7. Given the accuracy of your measurements have you verified Newton's 1st Law? Does the result meet your expectations? Discuss.

PART 3 ■

1. Repeat procedure described in **PART 1** but now use four strings and scales and an arrangement shown in figure 5-7.

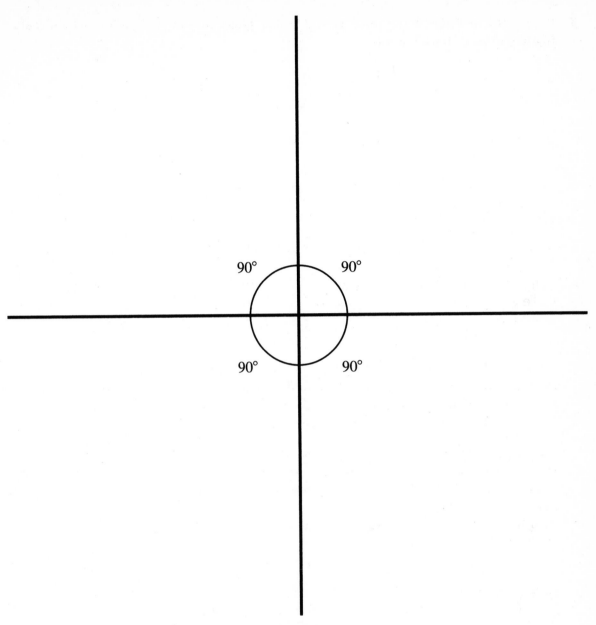

Figure 5-7.

2. Record the readings of the three spring scales. Make sure not to confuse which scale reading goes with each scale.

$$F_1 =$$

$$F_2 =$$

$$F_3 =$$

$$F_4 =$$

3. Do you notice anything particular about the magnitudes of these forces?

4. In figure 5-7 draw the horizontal forces acting on the ring. Assign a scale suitable to make a large clear drawing yet not run out of room (e.g. 1cm ~1N or 0.5 cm = 1N). Label the vectors.

5. Find the resultant force $F_{net\ 12}$ from adding F_1 and F_2 as well as $F_{net\ 34}$ from adding F_3 and F_4 (vectors F_1 and F_3 should lie along a single line and so should F_2 and F_4 if you labeled them accordingly)

$$F_{net\ 12} =$$

$$F_{net\ 34} =$$

How does the magnitude of $F_{net\ 12}$ compare with that of $F_{net\ 34}$?

How do their directions compare?

6. Do $F_{net\ 12}$ and $F_{net\ 34}$ balance each other?

7. Given the accuracy of your measurements have you verified Newton's 1st? Does the result meet your expectations? Discuss.

EXPERIMENT | SIX

Momentum Conservation

▣ Purpose

- To verify the principle of momentum conservation by observing balls recoiling on a ruler.

☑ Equipment and Supplies

- ❏ Grooved plastic ruler
- ❏ Scotch tape
- ❏ 3 identical steel balls and 1 brass, 1 aluminum, 1 glass, 1 wooden ball. The balls should be of similar size and small enough to roll and slide easily in the groove of the ruler
- ❏ Index cards
- ❏ Balance or electronic scale

INTRODUCTION

The momentum p of an object is defined as the product of its mass m and its velocity v, i.e.

$$p = mv \tag{1}$$

Like velocity momentum is a vector and has the same direction as the velocity vector. The principle of momentum conservation tells us that if the net external force acting on a system of objects is zero, then the total momentum of this system is conserved i.e. stays constant.

In this experiment you will verify the principle of momentum conservation by investigating the behavior of balls recoiling on a grooved ruler. If you place two balls of masses m_1 and m_2 which are initially at rest in the groove of the ruler positioned horizontally on the desk, and sandwich a folded index card in between them, the balls will recoil in opposite directions with speeds proportional to the inverse ratio of their masses after you release your finger grip on them. As we shall see, this rule is a consequence of momentum conservation, which here says that since the initial momentum P_i of the system is zero (since balls are at rest $v_{1i} = v_{2i} = 0$), their final momentum P_f must also be zero.

$$P_i = 0 = P_f, \tag{2}$$

where
$$P_f = m_1 v_{1f} - m_2 v_{2f}, \tag{3}$$

The "−" sign indicates that the velocity of the second ball points in opposite direction from that of the first ball. And so from the principle of momentum conservation we have $P_i = 0 = P_f = m_1 v_{1f} - m_2 v_{2f}$, and thus $m_1 v_{1f} = m_2 v_{2f}$, or

$$\frac{m_1}{m_2} = \frac{v_2}{v_1}. \tag{4}$$

which is the rule mentioned earlier. If we wish to estimate where the balls would be on the ruler at time t after collision, we can use the fact that the distance covered by the first ball is $D_1 = v_1 t$ and $D_2 = v_2 t$ and thus

$$\frac{m_1}{m_2} = \frac{D_2}{D_1}. \tag{5}$$

Now if we wish for the balls to arrive at the ends of the ruler simultaneously, where should we initially place the balls and the index card? Let D_{tot} denote the total length of the ruler, then if the balls reach the ends of the ruler at the same time

$$D_1 + D_2 = D_{tot}, \tag{6}$$

where we neglected the finite dimensions of the balls. Since the balls are the same size we can do that, and then regard the final result as a result for the position of the index card rather than the ball itself. From equations (6) and (5) we obtain

$$D_1 = \frac{D_{tot}}{\dfrac{m_1}{m_2} + 1}. \tag{7}$$

Since we can neglect the thickness of the folded index card in our calculations we can say that the index card should be positioned at D_1.

From equation (7) you can now predict where to place the two balls and the index card so that after the recoil the balls arrive at the ends of the ruler simultaneously, provided you know their masses.

PROCEDURE

PART 1 ■ TWO BALLS RECOILING ON A GROOVED RULER

1. Place the ruler horizontally on the desk, and tape it to the desk with scotch tape but be careful not to put any tape in the groove of the ruler.

2. Select your balls. You should have 2 identical steel balls, and 4 balls of different materials but similar radii. Place each ball on the scale and measure its mass. Record those values.

$$m_{steel\ 1} =$$

$$m_{steel\ 2} =$$

$$m_{brass} =$$

$$m_{aluminum} =$$

$$m_{glass} =$$

$$m_{wood} =$$

3. What is the total length of your ruler D_{tot}?

$$D_{tot} =$$

4. Use equation (7) to predict where the balls should be placed on the ruler so that after the recoil the balls arrive at the ends of the ruler simultaneously, and record your results in table 1.

TABLE 1

BALLS RECOILING	m_1 (g)	m_2 (g)	$D_{theoretical}$ (cm)
2 steel balls			
1 steel ball and 1 brass ball			
1 steel ball and 1 aluminum ball			
1 steel ball and 1 glass ball			
1 steel ball and 1 wooden ball			

5. Place each pair of balls (one pair at a time) at the predicted position on the groove of the ruler with the folded index card sandwiched between them (see figure 6-1). Squeeze the card closed by finger pressure on the balls. Suddenly and simultaneously remove your fingers from the balls and observe weather or not the balls reached the ends of the ruler at the same time. If the folded index card does not provide enough recoil for the balls to reach the ends of the ruler fold it second or even third time.

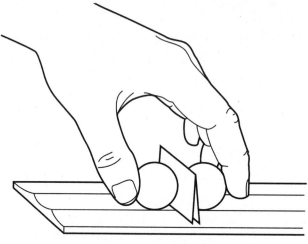

Figure 6-1. Apparatus.

6. Were your predictions made in table 1 verified experimentally for each pair of the balls? Discuss.

PART 2 ■ COLLISIONS

In this part you are asked to observe a series of collisions and then record your observations and try to analyze them in terms of physical phenomena that takes place, in particular the principle of momentum conservation.

1. Take two steel balls and measure their masses. Record the values in table 2 below. Then place them at the opposite ends of the grooved ruler. Launch the balls simultaneously towards each other with high speed by rapidly sweeping your hands together so the balls are sliding rather than rolling. Observe the balls before and after collision and fill table 2 with the results of your observations.

TABLE 2

		LEFT BALL	RIGHT BALL
MASS (g)			
Before collision	Speed*		
	Direction		
After collision	Speed*		
	Direction		

* indicate the speed of the balls with comparative statements like: faster, slower, the same or stationary.

Discuss the momenta of the balls before and after collision.

2. Take two steel balls and place one in the middle of the ruler and one at the end of the ruler. Launch the ball that sits at the end of the ruler towards the other ball with high speed by rapidly sweeping your hand so the ball is sliding rather than rolling. Observe the balls before and after collision and fill table 3 with the results of your observations.

TABLE 3

		LEFT BALL	RIGHT BALL
MASS (g)			
Before collision	Speed*		
	Direction		
After collision	Speed*		
	Direction		

* indicate the speed of the balls with comparative statements like: faster, slower, the same or stationary.

Discuss the momenta of the balls before and after collision.

3. Take a third steel ball and measure its mass. Record the masses of the balls in table 4. Place two balls in the middle of the ruler touching each other and one at the end of the ruler. Launch the ball that sits at the end of the ruler towards the other balls with high speed by rapidly sweeping your hand so the ball is sliding rather than rolling. Observe the balls before and after collision and fill table 4 with the results of your observations.

TABLE 4

		LEFT BALL	MIDDLE BALL	RIGHT BALL
MASS (g)				
Before collision	Speed*			
	Direction			
After collision	Speed*			
	Direction			

* indicate the speed of the balls with comparative statements like: faster, slower, the same or stationary.

Discuss the momenta of the balls before and after collision.

4. Take a steel ball and the ball with the smallest mass that you have available. Measure their masses and record them in table 5. Place the balls at the opposite ends of the grooved ruler. Launch the balls simultaneously towards each other with high speed by rapidly sweeping your hands together so the balls are sliding rather than rolling. Observe the balls before and after collision and fill table 5 with the results of your observations.

TABLE 5

		LEFT BALL	RIGHT BALL
MASS (g)			
Before collision	Speed*		
	Direction		
After collision	Speed*		
	Direction		

* indicate the speed of the balls with comparative statements like: faster, slower, the same or stationary.

Discuss the momenta of the balls before and after collision.

5. Take a steel ball and the ball with the smallest mass that you have available. Place the steel ball in the middle of the ruler and the light ball at the end of the ruler. Launch the light ball that sits at the end of the ruler towards the steel ball with high speed by rapidly sweeping your hand so the ball is sliding rather than rolling. Observe the balls before and after collision and fill table 6 with the results of your observations.

TABLE 6

		LEFT BALL	RIGHT BALL
MASS (g)			
Before collision	Speed*		
	Direction		
After collision	Speed*		
	Direction		

* indicate the speed of the balls with comparative statements like: faster, slower, the same or stationary.

Discuss the momenta of the balls before and after collision.

6. Take a steel ball and the ball with the smallest mass that you have available. Place the light ball in the middle of the ruler and the steel ball at the end of the ruler. Launch the steel ball that sits at the end of the ruler towards the light ball with high speed by rapidly sweeping your hand so the ball is sliding rather than rolling. Observe the balls before and after collision and fill table 7 with the results of your observations.

TABLE 7

		LEFT BALL	RIGHT BALL
MASS (g)			
Before collision	Speed*		
	Direction		
After collision	Speed*		
	Direction		

* indicate the speed of the balls with comparative statements like: faster, slower, the same or stationary.

Discuss the momenta of the balls before and after collision.

EXPERIMENT SEVEN

Torques

▮ Purpose

- To verify that for a body in equilibrium the sum of the torques must vanish.

☑ Equipment and Supplies

- ❏ Light weight ruler 30 cm (12 inches) long
- ❏ 24 pennies
- ❏ Scotch tape
- ❏ Round pen
- ❏ Balance or electronic scale

INTRODUCTION

In your daily experience you encounter not just the action of forces but also of torques. When you pull or push on a door knob to open an unlocked door you are applying torque to it; the door then swings on the hinges. As unbalanced force causes a change in the motion of an object, an unbalanced torque changes its rotational motion. Thus for an object to remain stationary (in equilibrium) not only forces but also torques acting on it need to be balanced, or add up to zero.

We define the torque as a product of a lever arm and the force applied to an object. The lever arm is the distance between the applied force and the rotational axis or fulcrum. We write

$$\tau = Fl, \tag{1}$$

where the Greek letter τ denotes the torque, F is the applied force, and l the lever arm.

From equation (1) you can see that the strength of the torque depends not only on the magnitude of the applied force but also on the length of the lever arm. Indeed you'd need to push the door with considerably grater strength for it to move if you apply this push close to the hinges rather then some distance away from them.

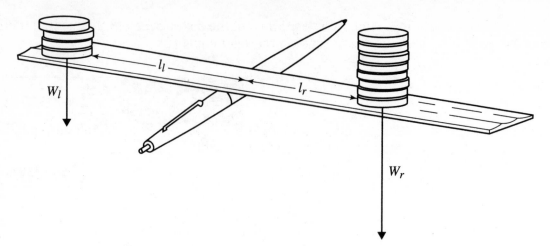

Figure 7-1. Pennies on the balanced ruler with pen as fulcrum.

Questions

a) If you double the length of the lever arm how much less force you need to exert to apply the same amount of torque to an object?

b) In what units is torque measured in the SI system?

The direction of rotation produced by a torque can be clockwise or counterclockwise. If you are on one side of a door, if you pull on the door knob the door will open, if you push on it, it will close. If you are on the other side of a door , if you pull on the door knob the door will close, if you push on it, it will open. Since torques can have such opposite effects you need to establish some kind of sign convention; for example all torques that produce clockwise rotation are positive and all that produce counterclockwise rotation are negative. If you pull on the door knob on one side of the door, and your friend pulls on it on the other side with equal strength, the door won't move—the torques the two of you produce are equal but have opposite signs since they would cause the door to rotate in the opposite direction and thus they cancel out producing no net effect.

In this lab you will put pennies on a ruler placed on a pen (see figure 7-1) and check if the torques in this system indeed cancel out when the ruler is balanced. Let us suppose for now that we can neglect the weight of the ruler itself. That leaves us with the weights of four pennies on the left and eight pennies on the right. Let us assume that each penny has mass m. Then on the left we have weight

$$W_l = 4m\,g,$$

and on the right

$$W_r = 8m\,g.$$

The eight pennies on the right are positioned at the distance l_r from the fulcrum (middle of the pen), and thus create a torque on the ruler equal to

$$\tau_r = W_r l_r = 8mg\, l_r$$

acting in the clockwise direction.

The four pennies on the left are positioned at the distance l_l from the fulcrum (middle of the pen), and thus create a torque on the ruler equal to

$$\tau_l = -W_l l_l = -4\, mg\, l_l$$

acting in the counterclockwise direction, thus the minus sign.

If the ruler is balanced it means that the two torques must add to zero:

$$\tau_l = \tau_r = 0.$$

Thus, $$8\, m\, gl_r - 4\, m\, gl_l = 0$$

and so $l_l = 2\, l_r$, or the ruler won't balance. You probably realized this was the relationship based on your experience with seesaws.

PROCEDURE

PART 1 ■

1. Place a pen on the desk and using scotch tape fasten its two ends to the desk so it can't roll around. Place the ruler perpendicularly on top of the middle of the pen with the centimeter scale facing you. Try balancing the ruler.

2. How long are the portions of the ruler on each side of the pen when the ruler is balanced? Locate the center of the ruler. (Note: the ruler may in fact be a little longer than 30 cm, since there usually are few millimeters at the ends before the scale starts and after it ends. These need to be considered.)

3. Stack eight pennies on top of each other and tape them together. Then place them on a scale and determine their mass. Record this value as m_r.

$$m_r =$$

4. Keeping the center of the ruler at the fulcrum tape the pennies to the middle of the right hand side of the ruler i.e. a quarter of the ruler's length from its right end. The center of the pennies should fall near the 22.5 cm mark. Record this value as l_r.

$$l_r =$$

5. Calculate the torque the 8 pennies placed at l_r exert on the ruler i.e. $\tau_r = W_r l_r = m_r g l_r$

$$\tau_r =$$

(Note: Pay attention to units)

6. Stack four pennies on top of each other and tape them together. Then place them on a scale and determine their mass. Record this value as m_l in table 1.

TABLE 1

NUMBER OF PENNIES ON THE LEFT	m_l (kg)	l_l (m)	τ_l (Nm)	τ_r (Nm)	% DIFFERENCE
4					
6					
8					
10					
12					
14					
16					

7. Keeping the center of the ruler at the fulcrum place the four pennies on the left hand side of the ruler. By moving the 4 pennies towards and/or away from the fulcrum find the location at which the four pennies on the left balance the eight pennies on the right. The distance should be measured from the middle of the fulcrum to the center of the pennies. Record this value as l_l in table 1.

8. Repeat steps 6 and 7 with 6, 8, 10, 12, 14, and 16 pennies balanced on the left hand side of the ruler. Leave the 8 pennies taped to the middle of the right hand side of the ruler. You'll need them there for Part 2.

9. Calculate the torques which the pennies on the left hand side of the fulcrum exert on the ruler and record them in table 1. (Note: Pay attention to units).

10. If the ruler is balanced then the two torques should be equal in magnitude and oppositely directed. Calculate the percent difference between the magnitudes of two torques and record it in table 1. Is the percent difference between the magnitudes of two torques reasonably small? Discuss your results.

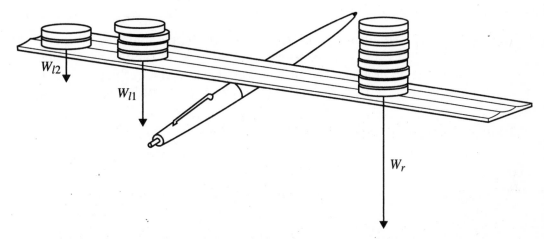

Figure 7-2. Three stacks of pennies on the balanced ruler with pen as fulcrum.

PART 2 ■

1. As in Part 1 have the ruler's center at the fulcrum and the eight pennies taped to the middle of the right hand side of the ruler.

2. Stack four pennies on top of each other and tape them together. Then place them on a scale and determine their mass as you did in Part 1. Record this value as m_{l1}.

$$m_{l1} =$$

3. Keeping the center of the ruler at the fulcrum tape the four pennies to the middle of the left hand side of the ruler i.e. a quarter of the ruler's length from its left end. The center of the pennies should fall near the 7.5 cm mark. Record this value as l_{l1}

$$l_{l1} =$$

4. Calculate the torque the 4 pennies placed at l_{l1} exert on the ruler.

$$\tau_{l1} =$$

5. Pick up two more pennies and measure their mass. Record this value.

$$m_{l2} =$$

6. Keeping the center of the ruler at the fulcrum place the two pennies on the left hand side of the ruler. By moving them (stacked together) towards and/or away from the fulcrum find the location at which the ruler is balanced. Measure the distance from the middle of the fulcrum to the center of the two pennies. Record this value as l_{l2}.

$$l_{l2} =$$

7. Calculate the torque the two pennies placed at l_{l2} exert on the ruler.

$$\tau_{l2} =$$

8. Calculate the torque which all the pennies on the left hand side of the fulcrum exert on the ruler.

$$\tau_l =$$

9. Within experimental accuracy is the total torque exerted on the left hand side the same in magnitude as that exerted on the right hand side of the ruler? (You found it in Part 1) Calculate the percent difference between the magnitudes of these two torques. Discuss your results.

EXPERIMENT | EIGHT

Work and Energy Conservation

Purpose

- To verify the law of energy conservation by launching an object upward and comparing its maximum height with the theoretical prediction.
- To verify that work is equal to force times displacement by pulling an object up an incline.
- To verify that work is equal the difference between initial and final energies by pulling an object up an incline.

Equipment and Supplies

- ☐ Meter stick
- ☐ Scotch tape
- ☐ Retractable ballpoint pen which part can be launched by the compressed spring
- ☐ Balance or electronic scale
- ☐ Steel ball (about 0.5 kg) outfitted with a hook
- ☐ Spring scale
- ☐ Wooden wedge or block
- ☐ Large hardcover plastic coated book
- ☐ Protractor

INTRODUCTION

We define the work done by a force F as the product of that force and the displacement d it caused (we assume that the force is constant and that the motion takes place in the direction of the force)

$$W = Fd. \tag{1}$$

Thus the amount of work you do to lift your bag off the floor and onto the desk is equal to the amount of force you exert on the bag times the height of the desk you lifted it to.

The conservation of energy principle tells us that energy of an isolated system of objects cannot be created or destroyed, it can only be converted from one form to another.

Thus, in the absence of forces such as friction that add or remove energy from a system, the mechanical energy (kinetic plus potential) remains constant. In this experiment the potential energy of a compressed spring in a ballpoint retractable pen will be converted to a kinetic energy of a launched top part of the pen, then as this part is projected upward into gravitational potential energy, to be finally converted back to kinetic energy on its way down.

The potential energy of the compressed spring is given by the following expression:

$$PE = \frac{1}{2} kL^2, \tag{2}$$

where PE denotes the potential energy, L is the distance that the spring is compressed from its equilibrium position, and k is the spring constant.

The kinetic energy KE of an object of mass m moving with a velocity v is given by:

$$KE = \frac{1}{2} mv^2. \tag{3}$$

The gravitational potential energy is given by

$$PE = mgh, \tag{4}$$

where m is the mass of an object, h is the height, and $g = 9.8 m/s^2$ is the gravitational acceleration on Earth.

To confirm the principle of energy conservation we will compare the potential energy of a fully compressed spring to the gravitational potential energy of the top part of the pen at its maximum height, assuming that the pen was launched straight up. At that point the top part of the pen is momentarily stationary and so its kinetic energy is zero. We have, based on conservation of energy:

$$\frac{1}{2} kL^2 = mgh, \tag{5}$$

and thus the height to which the top part of the pen should rise is

$$h = \frac{kL^2}{2mg}. \tag{6}$$

We can therefore predict how high the top of the pen should rise providing we know the spring constant k. To find k we measure how much force is needed to fully compress the spring by a distance L

$$F = kL. \tag{7}$$

F is found by placing the spring on a scale and measuring the amount of mass M needed to balance the force you exert to fully compress the spring. Then this force is equal to

$$F = Mg, \tag{8}$$

and so from equations (7) and (8), we can finally find

$$k = \frac{Mg}{L}.$$ (9)

Combining equations (6) and (9) we find the theoretical estimate for the maximum height based on conservation of energy

$$h = \frac{ML}{2m}.$$ (10)

PROCEDURE

PART 1 ■ ENERGY CONSERVATION

1. Disassemble the pen, but leave the spring on the top of the ink-carrying stem, which should be held vertically with its bottom end on your desk. Place the front plastic part of the pen on top of the spring, and compress the spring by grasping the lower end of the plastic part between your thumb and index finger, and pulling it down as far as it will go (see figure 8-1).

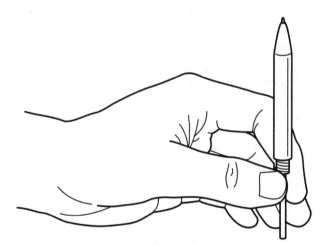

Figure 8-1. Experimental setup for PART 1.

Suddenly release your finger grip on the bottom of the plastic part and observe as it jumps in the air. Practice shooting the top plastic part vertically into the air.

2. Using the scale measure the mass m of the launched plastic part of the pen. Record this value.

$$m =$$

3. Place the spring on the scale and *barely* compress it fully. If you are using an electronic scale it will give you a reading of the mass M. If you are using balance scale place enough mass on the other pan to achieve the balance—that is your value of M. Record this value.

$$M =$$

4. With a ruler measure the distance L by which the spring compresses. Record this value.

$$L =$$

5. Estimate the theoretical value for the maximum height to which the top plastic part of the pen should ascend using formula (10).

$$h_{\text{theoretical}} =$$

6. To get a good estimate on how high the top plastic part of the pen ascends place a vertical meter stick next to the pen. Have your partner keep his or her head at the anticipated maximum height and launch the pen so that its distance from his or her face is the same as the meter stick, you should be able to estimate the height with an uncertainty of a centimeter. Practice shooting the pen and estimating the maximum height several times. Remember to account for the height of the top plastic part of the pen itself in your measurements.

7. Measure the maximum height to which the top plastic part of the pen ascends five times. Use the highest of the five values as your best result. Record this value of h.

$$h_{\text{experimental}} =$$

8. Compare the experimental h value with the theoretical prediction by calculating the percent difference. Are your results in good agreement considering the accuracy of your measurements? Discuss.

9. Discuss how and when different forms of energy are converted to one another during this experiment.

PART 2 ■ WORK

1. Construct an incline by placing a book on a wedge or a wooden block as shown in figure 8-2. Tape the wedge or block to the desk so it will not slide. Use a protractor to ensure the inclined plane of the book makes no more than a 15° angle with the horizontal.

2. Place a metal ball outfitted with a hook on the bottom of the incline with the hook side up as shown in figure 8-2. The ball should not stick out beyond the incline.

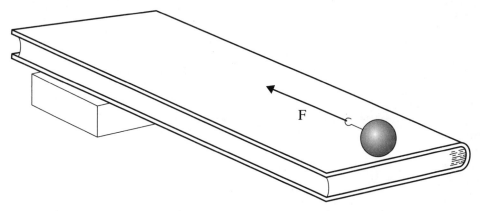

Figure 8-2. Experimental setup for PART 2.

3. Measure the vertical height h_1 of the middle of the ball in reference to the desk's surface. Record this value.

$$h_1 =$$

4. Now place the ball on top of the incline. The ball should not stick out beyond the incline (the hook should, though). Measure the vertical height h_2 of the middle of ball in reference to the desk's surface. Record this value.

$$h_2 =$$

5. Place the ball on a scale and measure its mass m. Record this value.

$$m =$$

6. Place the ball back on the bottom of the incline and attach a spring scale to the hook. With a steady pull drag the ball up the incline to its highest possible position (without the block sticking out beyond the incline). Use the minimum amount of force needed to move the ball. The scale should show a steady reading for the force (remember to multiply the value shown by the spring scale by g to get the force). Repeat the measurement three times and record the average value of the force.

measurement # 1 $F =$

measurement # 2 $F =$

measurement # 3 $F =$

$$F_{ave} =$$

7. With a ruler measure the distance d you moved the ball up the incline. Record this value.

$$d =$$

8. Using equation (1) estimate the amount of work you used to pull the ball up the incline.

$$W =$$

9. What force/forces were you working against while pulling the ball up the incline?

10. The work done against gravity is equal to the difference between the final and initial potential energies of the ball. Using equation (4) and the values for final and initial height of the ball above the desk which you measured in **PART 2.3** and **2.4** calculate the final and initial potential energies of the ball. Then calculate the work done against the gravity by computing the difference between the final and initial potential energies of the ball.

$$PE_i =$$

$$PE_f =$$

$$W =$$

11. Is the value you obtained in **PART 2.10** for work similar to that obtained in **PART 2.8.** Why not?

12. To account for the work done to overcome friction between the ball and books surface we need to measure the frictional force. Take the wedge or the wooden block out from underneath the book and place the book horizontally on the table. With the spring scale steadily pull the ball across the book and measure the force needed to move it just as you did with the book in the inclined position. Use the minimum amount of force needed to move the ball. Remember to multiply the value shown by the spring scale by g to get the force. Repeat the measurement three times and average your results. Record the average value of the frictional force F_f.

measurement # 1 $F_f =$

measurement # 2 $F_f =$

measurement # 3 $F_f =$

$F_{f \, ave} =$

13. Using formula (1) and the value for F_f which you have just found calculate how much work you did to overcome friction while you pulled the block with the spring scale.

$W_f =$

14. The work you did pulling the block up the incline (calculated in **PART 2.8**) was a total of the work against gravity which you did to lift the block form height h_1 to height h_2 (calculated in **PART 2.10**), and the work you did to overcome friction (calculated in **PART 2.13**). Thus if you add these two values you should obtain a result close to the total work done calculated in **PART 2.8.** Calculate the total work done by adding the work done against gravity and the work done to overcome friction. Is your result similar to that obtained in **PART 2.8**? Compare these two values by calculating percent difference. Discuss your results.

$W_{tot} =$

EXPERIMENT | **NINE**

Archimedes' Principle

◨ Purpose

- To verify Archimedes' principle by measuring the weight of water displaced by an object submerged in it; to predict based on their density whether objects will float or sink.

☑ Equipment and Supplies

- ❏ A steel ball about 5 cm in diameter, a golf ball, a solid rubber ball, a wooden block, and a styrofoam block. The steel ball should be outfitted with a hook; the wooden block should have a scale in millimeters glued to its shortest side, and be coated with plastic polymer so as not to get waterlogged.
- ❏ Ruler
- ❏ Caliper
- ❏ Calibrated cylinder
- ❏ Spring scale
- ❏ Plastic ice cream bowl
- ❏ Styrofoam cup
- ❏ Helium balloon
- ❏ Ball of heavy string
- ❏ Balance or electronic scale
- ❏ Large plastic container

INTRODUCTION

Archimedes' principle states that an object placed in a fluid experiences an upward buoyant force acting on it. This buoyant force is equal to the weight of the fluid displaced by the object. The principle applies to all fluids, both liquids and gases.

If the density of an object is less than the density of the fluid it is placed in, then the object floats. A floating object is submerged in the fluid to a depth that is sufficient to displace the weight of the fluid equal to the object's weight. The object is then in equilibrium under the action of two forces of equal magnitude: its weight acting downward, and a buoyant force acting upward. If the buoyant force is less than the weight of the object, the object sinks.

The density ρ of an object is defined as the mass in of the object m divided by its volume V. In equation form this is

$$\rho = \frac{m}{V} \tag{1}$$

The metric units for density are kg/m³. Another commonly used unit for density is g/cm³ (grams per cubic centimeter). The density of water is 1.000 g/cm³.

a)   b)

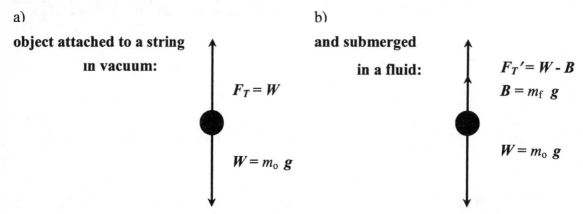

Figure 9-1. Object suspended by a string in vacuum (a) and submerged in a fluid (b).

For objects whose density is greater than the density of a fluid, Archimedes' principle allows a rather simple determination of the density of the object. Consider an object that is tied to a string attached to a spring scale, as shown in the figure 9-1a. In figure 9-1a you see a free body diagram of an object suspended by a string in vacuum (or air, since most objects are heavy enough that the buoyancy of air can be neglected). The object is not moving i.e. it is in the state of equilibrium. This means that the weight of the object $W = m_o \boldsymbol{g}$ is exactly balanced by the force of tension in the string. So $F_T = W = m_o \boldsymbol{g}$. Now let's consider what happens when this object is submerged in a fluid. The object again is in equilibrium. Gravity is still acting on this object with the same strength, i.e. it still weighs $W = m_o \boldsymbol{g}$, but now the fluid is exerting an upward buoyant force $\boldsymbol{B}$ on the object. According to the Archimedes' principle this buoyant force is equal to the weight of fluid displaced by the object $B = m_f \boldsymbol{g}$. Thus the tension F_T' in the string which is the apparent weight of the object submerged in the fluid, is reduced as compared with F_T. Specifically,

$$F_T' = W - B \tag{2}$$

Since the mass of the displaced fluid is the product of its density and volume we can write

$$B = m_f \boldsymbol{g} = \rho_f V_f \boldsymbol{g}, \tag{3}$$

where ρ_f and V_f are the density and the volume of the displaced fluid respectively.

If the object is entirely submerged in the fluid then it displaces the volume of fluid equal to its own volume, and thus $V_f = V_o$, where V_o denotes the volume of the object. Equation (3) becomes

$$B = m_f \boldsymbol{g} = \rho_f V_f \boldsymbol{g} = \rho_f V_o \boldsymbol{g} \tag{4}$$

By combining Eq. (2) and (4) we can derive a following equation

$$\rho_o = \frac{W}{W - F_T'} \, \rho_f \tag{5}$$

Equation (5) implies that the density of any object that sinks in fluid can be determined by measuring its weight W and its apparent weight F_T' in fluid. This experiment will make use of equation (5) to determine the density of an object submerged in a fluid.

    You have to use a different procedure to determine the density of an object that floats on the surface of the fluid (now the object is no longer suspended by a string). The volume of the fluid displaced is less than the volume of the object that is floating since only a portion of the object is under the fluid's surface, thus $V_f \neq V_o$. The buoyant force is now completely supporting the object (see figure 9-2),

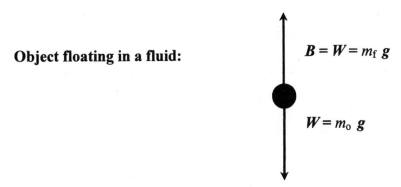

Object floating in a fluid:

$$B = W = m_f \, g$$

$$W = m_o \, g$$

Figure 9-2. Object floating on the surface of the fluid.

so $B = m_f \, g = \rho_f \, V_f \, g = W = \rho_o \, V_o \, g$. Therefore we find

$$\rho_o = \frac{V_f}{V_o} \cdot \rho_f \tag{6}$$

In other words, the density of the object as a fraction of the density of the fluid just equals the fraction of the object's volume that is submerged. In this experiment you will estimate V_f and V_o and determine the density of the object.

PROCEDURE

PART 1 ■ DENSITY

1. Examine the three balls and three blocks in front of you and make predictions which are going to float and which will sink in water. Record your predictions in table 1.

TABLE 1

OBJECT	ORIGINAL PREDICTION F-FLOATS S-SINKS	MASS (g)	VOLUME (cm³)	DENSITY (g/cm³)	BASED ON ITS DENSITY WILL THE OBJECT FLOAT OR SINK IN WATER?	WAS YOUR ORIGINAL PREDICTION TRUE OR FALSE?
Steel ball						
Golf ball						
Rubber ball						
Wooden block						
Styrofoam block						

2. Use the electronic or balance scale to determine the mass of each of the balls and of each of the blocks. Record these masses in grams as m in table 1. Then measure the dimensions of these objects (you need to measure the diameter of each ball and length, width, and height of each block). Calculate the volume of each object using the formulas below:

$$V_{block} = (length)\ (width)\ (height)$$

$$V_{ball} = (4\ /\ 3)\ \pi\ (diameter\ /\ 2)^3 \qquad (\pi = 3.14),$$

and record the volumes in table 1. Then calculate the density of each object in grams per cubic centimeter (g/cm³) using equation (1) and record it in table 1. Compare these densities with the density of water and determine which objects will float and which will sink in water. Discuss your results.

3. Place the objects in a container half filled with water and observe whether they sink or float.

4. Place an empty calibrated cylinder on the electronic or balance scale. Record its mass. Then pour 100 milliliters (ml) of water into the calibrated cylinder and measure the mass. Determine the mass and the density of water in the cylinder.

Mass of an empty **calibrated cylinder** . . . .   _____

Mass of a **calibrated cylinder** containing 100 ml of water . . . .   _____

Mass of 100 ml of water . . . .   _____

Volume of 100 ml water in cm³ . . . .   _____

Density of water in g/cm³ is . . . .   _____

Compare with the accepted value of 1 g/cm³ (calculate % deviation)

Is your calculated value for the density of water close to the accepted value of 1 g/cm³? Discuss.

PART 2 ■ ARCHIMEDES' PRINCIPLE—STEEL BALL IN WATER

1. Using the electronic or balance scale measure the mass of the empty ice cream bowl. Record this value.

$$m_{\text{empty bowl}} =$$

2. Fill the styrofoam cup with water all the way to the brim and place it in the ice cream bowl.

3. Suspend the metal ball from the metal hook on the bottom of the spring scale and hold it steady vertically (see figure 9-3).

Figure 9-3. Ball suspended by a spring scale in air.

From the reading of the scale calculate the tension force F_T, which is now fully balanced by the weight W of the ball (see figure 9-1a).

$$F_T = W =$$

4. Carefully insert the ball into the water until it is fully submerged. Make sure the ball is not touching the bottom or the sides of the cup and that it is supported solely by the spring scale (see figure 9-4).

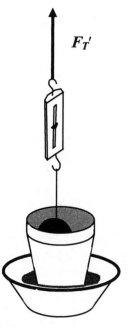

Figure 9-4. Ball suspended by a spring scale and submerged in water.

Record the tension force for the ball fully submerged in water. This is the apparent weight F_T' of the ball fully submerged in water.

$$F_T' =$$

5. Using electronic or balance scale measure the mass of the ice cream bowl with the displaced water. Record this reading.

$$m_{\text{bowl with water}} =$$

Calculate the mass of the displaced water and its weight.

$$m_{\text{water}} =$$

$$W_{\text{water}} =$$

Since according to Archimedes' principle the weight of the displaced water is equal to the buoyant force with which the water acts on the ball determine the buoyant force **B**.

$$B =$$

6. Add the values of tension force F_T' and buoyant force **B** together and compare the result with the weight **W** of the ball which you measured in **PART 2.3.** Are they the same considering the accuracy of your measurements? Explain why.

$$B + F_T' =$$

7. Use equation (5) to determine the density of the metal ball. Is this estimate of density of the same order in magnitude as the value you calculated in **PART 1**? Explain.

PART 3 ■ DENSITY OF A FLOATING OBJECT

1. Half-fill the large container with water. Carefully place the wooden blocks so that they float in the water. Use the scale attached to the blocks to estimate the fraction of the block that is under water (V_f/V_o) and calculate the density of the wooden block using equation (6).

Measure the fraction of the wooden block that is under water . . . . . .   _____

Calculate . . . . $V_f/V_o =$ _____

The density of the wooden block . . . . $\rho_o =$ _____

Is this estimate of density approximately the same as the value you calculated in **PART 1** ? Explain.

PART 4 ■ **BUOYANCY OF HELIUM BALLOONS**

1. Measure the force required to hold a helium balloon stationary by tying the string to the balloon. The balloon should not be inflated too much. Tie one end of the string to the balloon, place the ball of string on the desk and let enough of the string unwind so that the balloon floats without rising or falling. Cut the string at the spool and verify that the balloon remains stationary. If not adjust the length of the string by cutting some of or tying in more, to where balloon neither falls nor rises.

2. Untie the string from the balloon and using electronic or balance scale measure its mass.

    Mass attached to the balloon needed to keep it stationary is . . . . _____

    The force required to hold the balloon stationary is . . . . _____

    What makes the balloons float in air? Discuss forces involved.

3. Why were we able to neglect the buoyancy of air in **PART 2**, but we cannot here in **PART 4**?

TABLE 2			
TIME (MIN)	TEMPERATURE T OF THE WATER IN THE GLASS (°C)	THE EXCESS ABOVE ROOM TEMPERATURE T_s (°C)	THE % BY WHICH THE TEMPERATURE DECLINED IN THE LAST MINUTE
0			
1			
2			
3			
4			
5			
6			
7			
8			
9			
10			
11			
12			
13			
14			
15			
16			
17			
18			
19			
20			
21			
22			
23			
24			
25			

7. Measure the temperature of the water inside the glass every minute. Have one person monitor the stopwatch and another one take readings of the temperature of the water inside the glass. Record your results in table 2.

8. Continue acquiring data for 25 minutes.

9. In the third column of table 1 calculate the excess above room temperature (by subtracting T_{room} from the temperature values in the second column).

10. In the fourth column of table 1 calculate the percentage by which the excess above the environment temperature declined during the preceding minute i.e.

$$\frac{T_e(t = 0 \text{ min}) - T_e(t = 1 \text{ min})}{T_e(t = 0 \text{ min})} \times 100\% \text{ for the first minute and so forth.}$$

11. Plot the excess above room temperature of water as a function of time on a computer or in the space provided in figure 10-2. Remember to label your graph.

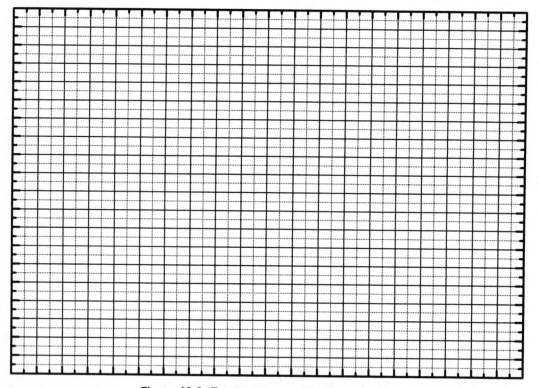

Figure 10-2. Temperature excess versus time.

12. Does the temperature excess above room temperature decline by the same percent each minute? Discuss.

13. How much time ($t_{1/2}$) does it take the water to cool down to $\frac{1}{2}$ of its original excess above room temperature? How much time ($t_{1/4}$) does it take the water to cool down to 1/4th of its original excess above room temperature?

$$t_{1/2} =$$

$$t_{1/4} =$$

14. Predict how much time ($t_{1/8}$) would it take the water to cool down to 1/8th of its original excess above room temperature? Predict how much time ($t_{1/16}$) would it take the water to cool down to 1/16th of its original excess above room temperature? Predict how much time ($t_{1/32}$) would it take for the excess above room temperature to become 1/32th of its original value? How are these times $t_{1/2}$, $t_{1/4}$, and $t_{1/32}$ related?

$$t_{1/8} =$$

$$t_{1/16} =$$

$$t_{1/32} =$$

15. How would the results of your experiment change if instead of 25 milliliters you used a liter of water?

16. Why is it natural that a hot object approaches the temperature of its surroundings at faster rate at first than later on rather than cooling at the constant rate all the time?

17. What are the plausible modes of energy transfer between hot water and its surroundings?

Simple Harmonic Motion

▣ Purpose

- To investigate simple harmonic motion of a pendulum.

☑ Equipment and Supplies

- ❏ A steel ball about 5 cm in diameter outfitted with a hook
- ❏ Set of weights
- ❏ Protractor
- ❏ String
- ❏ Spring
- ❏ Stable mount
- ❏ Ruler
- ❏ Stopwatch

INTRODUCTION

If an object moves under the action of a force which is proportional and directed opposite to its displacement, the object undergoes simple harmonic motion when displaced from equilibrium. The time it takes such an object to return to the initial position moving in the same direction is called the period. In simple harmonic motion, the period is independent of the maximum displacement from the equilibrium position known as the amplitude of the motion.

A bob of mass m suspended on a thin, light string of length L is a good approximation of a simple pendulum. Let's consider the forces acting on the bob. When hanging vertically the weight of the bob $W = mg$ is exactly balanced by the tension force of the string F_T (see figure 11-1a). However, when the bob is displaced until string makes an angle θ with the vertical and released, the tension force does not completely balance the weight since these two forces do not act along a single direction anymore (see figure 11-1b). The weight still acts vertically downwards, but the tension of the string F_T acting along the string now makes angle θ with the vertical. If we consider components of the weight: one acting along the string direction ($W_\parallel$) and one perpendicular to it ($W_\perp$), then we can see that the component along the string ($W_\parallel$) is exactly balanced by the tension of the string F_T, thus there is no motion in this direction (the string does not stretch or contract), but the component of the weight perpendicular to the string ($W_\perp$) remains unbalanced, thus being a source of acceleration when the bob is released. Thus the bob accelerates but with a decreasing rate since as the angle θ, which it makes with the vertical, decreases, so does the component of the weight perpendicular to the string ($W_\perp$). Eventually the bob reaches its equilibrium position, but since it already gathered a lot of speed, it doesn't stop there but continues to move. Through the principle of energy conservation we can relate the maximum height it achieves to its maximum speed

$$mgh_{max} = \frac{mv^2_{max}}{2}.$$

The period T of the pendulum is the time for one complete oscillation of the system. The amplitude of the oscillations (i.e. the maximum displacement of the pendulum) is related to the maximum angle θ which the bob attains with the vertical. In this experiment, you will investigate whether and how the period of oscillations of the pendulum depends on its mass, length, and the amplitude (or maximum angle θ with the vertical) of oscillations.

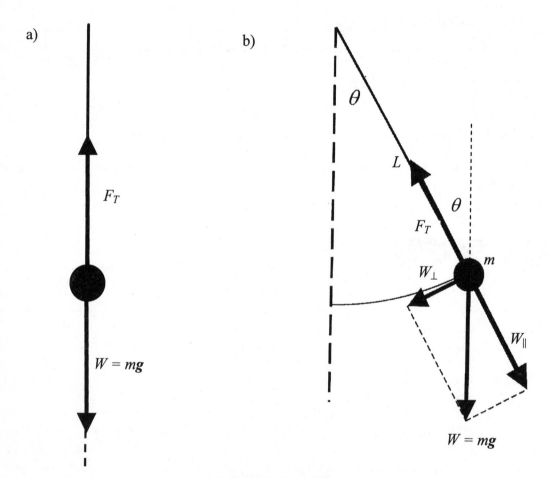

Figure 11-1. Pendulum.

PROCEDURE

PART 1 ■

1. Construct the pendulum by suspending the metal ball from a string so that its length (from the middle of the support to the center of the ball) is approximately 1m. Attach the top end of the string to a stable mount.

2. Displace the mass and release it. The pendulum will swing or oscillate.

3. Using a protractor, start the pendulum with a 7° angular amplitude, release it, and measure the period of its oscillations. Take 10 measurements. Tabulate the data in table 1 and calculate the average period.

TABLE 1

MEASUREMENT NUMBER	T(s)
1	
2	
3	
4	
5	
6	
7	
8	
9	
10	
The average period is	

4. Now time 10 full oscillations. Calculate the period by dividing this time by ten.

$$t_{10 \text{ oscillations}} =$$

$$T =$$

5. Is result from **PART 1.2** or **PART 1.3** more accurate? Why?

6. What factors limit the accuracy of your measurements?

7. Through discussions with other student groups estimate how accurately can you measure the time ten full oscillations take.

8. How accurately, therefore, can you determine the period?

PART 2 ■

1. Using a protractor to measure the angle between the string and the vertical, start the pendulum at 5 different small angles (15° or less) and measure the time of 10 full oscillations for each angle.

2. Collect your data in table 2 and calculate the period for each angle θ.

TABLE 2		
ANGLE θ (°)	**10 T (s)**	**T (s)**

3. Considering the accuracy of your measurements would you say that the period depends or does not depend on amplitude? Does this result meet your expectations? Discuss.

PART 3 ■

1. Time 10 full oscillations for 5 different masses of the bob. Use the bobs of 1kg, 0.5 kg, 0.2 kg, 0.1 kg, 0.05 kg. **NOTE: BE CAREFUL NOT TO DROP THE MASSES ON YOUR TOES.** Make sure you keep the length of the pendulum fixed (Note: the heights of these different weights may vary. You need to adjust the length of the string so that the length of your pendulum from the middle of the support to the center of the weight remains 1m for each weight).

2. Collect your data in table 3 and calculate the period for each mass.

TABLE 3

MASS (kg)	10 T (s)	T (s)

3. Considering the accuracy of your measurements would you say that the period does not depend on mass? Does this result meet your expectations? Discuss.

PART 4 ■

1. Using the metal ball as a bob, time 10 full oscillations for 5 different lengths of the pendulum. The lengths your group should use should span the whole range from 15 cm to the maximum length your equipment allows.

2. Collect your data in table 4 and calculate the period of the pendulum for each length.

TABLE 4

LENGTH (m)	10 T (s)	T (s)

3. What period do you expect for zero length? Include that value in your table.

4. In figure 11-2 or on your computer plot T versus of length. Is your data linear? Can you fit it well with a straight line?

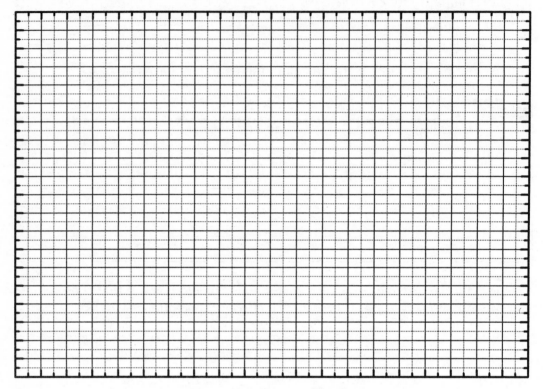

Figure 11-2. *T* **versus length.**

5. In table 5 calculate the square of the period of the pendulum.

LENGTH (m)	10 T (s)	T^2 (s)

6. In figure 11-3 or on your computer plot T^2 versus of length. Is your data linear? Can you fit it well with a straight line? Is this a better fit than you found when plotting T versus L?

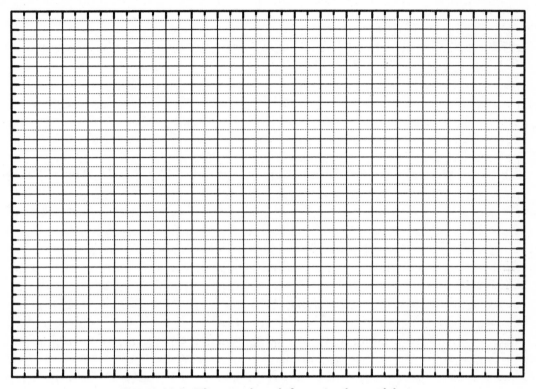

Figure 11-3. T^2 versus length for a simple pendulum.

7. Identify the slope of the fitted line. You can do this on your computer. If you do not have the computer available to you then on your graph (figure 11-3) draw a straight line with a ruler with trying to match your data the best you can. Then calculate the slope of this line: choose two points lying on the line as far apart as possible and find their coordinates l_1, T_1^2 and l_2, T_2^2, then calculate the differences: $\Delta T^2 = T_2^2 - T_1^2$, and $\Delta l = l_2 - l_1$, and find the slope from equation

$$slope = \frac{\Delta T^2}{\Delta l}.$$

8. According to your data what is the relation between the period and the length?

9. The actual formula for the period of the pendulum is

$$T = 2\pi\sqrt{\frac{L}{g}}. \tag{1}$$

where $\pi \approx 3.14$ and $g = 9.8$ m/s^2 is the acceleration due to gravity. If we square both sides of this equation we obtain

$$T^2 = \frac{4\pi^2}{g}L, \tag{2}$$

thus your fitted slope should be $\dfrac{4\pi^2}{g}$. Calculate this value using the accepted value of g.

$$\frac{4\pi^2}{g} =$$

10. Compare your experimental and theoretical values for the slope by computing the percent difference. Discuss your results.

11. If you took your pendulum to the Moon, where the acceleration due to gravity is one sixth of that on Earth, would you expect the period of the same pendulum to be longer, shorter, or the same as on Earth? By how much? Explain why.

Waves

◼ Purpose

- To investigate wave phenomena on a slinky.

☑ Equipment and Supplies

- ❏ Slinky
- ❏ Stopwatch
- ❏ Measuring tape (25 ft or more)
- ❏ Meter stick
- ❏ Chalk
- ❏ String
- ❏ Red ribbon
- ❏ Spring scale
- ❏ Electronic or balance scale

INTRODUCTION

Most of us are quite familiar with wave phenomena since waves abound in our lives: we all have observed waves on the surface of water, heard sound waves, seen light (electromagnetic waves), to name just a few. All these waves have a source in a vibrating object, and regardless of the type of wave share some common characteristics. You will examine some of these characteristics in this lab by investigating mechanical waves produced with a slinky.

Waves carry energy from one place to the other but at the same time the medium through which they propagate, in this case a slinky, does not move as a whole: the wave travels through a slinky but each piece of the slinky only vibrates about its equilibrium position.

Lets first consider a single pulse such as can be formed in a slinky or a rope by a sudden sideways jerk. The hand moves the end of the slinky up, and since the end displaced piece is attached to the adjacent pieces it in turn pulls them to the same side and so they also begin to move. As each subsequent piece of the slinky moves up a pulse (or a wave crest) travels along the slinky, while the end piece has returned to its original position by the motion of a hand. Thus an abrupt disturbance—a jerk is a source of the wave and the cohesive forces between adjacent pieces of a slinky cause the pulse to propagate forward. If the disturbance is smooth, continuous and periodic—such as a hand jerking the end of the slinky left and right at a uniform pace, a continuous wave is formed.

Let us define some important characteristics of such a wave created by vertical vibrations in a string (see figure 12-1). We call the high points of the wave the crests and the low points the troughs. The maximum displacement of a piece of the string from its equilibrium position i.e. the height of a crest or a depth of a trough is called the wave's amplitude (A). The distance between two successive crests (or troughs) defines the wavelength λ of the wave. The frequency is defined as a number of crests (or troughs) that pass a given point on the string per unit time. The period is defined as the time it takes one complete oscillation (e.g. from crest to crest) to pass a given point on its path. Thus $T = 1/f$. The wave velocity is the velocity with which the crests move along the rope and so

$$v = \lambda f. \tag{1}$$

For a wave propagating in a rope or spring stretched with tension force F the speed of the wave travelling in it is

$$v = \sqrt{\frac{F}{\mu}} \tag{2}$$

where μ is the linear mass density i.e. mass per unit length of the rope or spring which can be found by dividing the total mass of the rope or spring by its length:

$$\mu = m/L. \tag{3}$$

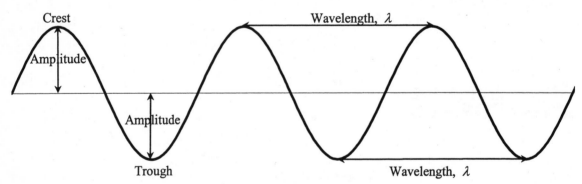

Figure 12-1. Characteristics of a wave.

An important distinction is made between the longitudinal waves in which the oscillations occur in the same direction as that in which the wave propagates ("back-and-forth"), and the transverse waves in which the oscillations occur in the direction perpendicular to the wave propagation ("side-to-side"). Examples of longitudinal waves include compressing and stretching a taut slinky and sound waves. In such a wave we call the crests and troughs compressions and rarefactions, respectively. Examples of transverse waves include wave travelling down a rope (or slinky) generated by moving your hands left and right perpendicularly to the rope (or slinky), and light waves.

When two or more waves travel in the same medium at the same time, the resultant wave can be found by adding the displacements produced by individual waves at each point. As a result, two or more travelling waves can pass through each other without being destroyed or altered just like the ripples produced by two stones thrown into water some distance apart do—we say they form an interference pattern. If the crests of two waves overlap then they produce a wave of increased amplitude—we call it constructive interference. If on the other hand, a crest meets a trough, the resulting wave's amplitude is reduced—we call it destructive interference.

When a wave strikes an obstacle or comes to the end of the medium e.g. the opposite end of the rope tied to the door knob, at least part of the wave is reflected back along the medium. You can observe this when a ripple on the surface of the water hits a rock. If the wave encounters a boundary between two media in which it can travel, some of it may be reflected back, and some transmitted to the second medium. The incident and reflected waves travel in the same medium in opposite directions and interfere with each other. This usually results in a complex pattern. But if the frequency is just right, the incident and reflected waves can interfere in such way that a standing wave is produced (it is called standing because it does not appear to move along the medium). In a standing wave parts of the pattern called the nodes are completely stationary – their displacement is zero due to destructive interference. The parts of the pattern that oscillate with maximum amplitude are called antinodes and they occur half way between the nodes.

Consider a standing wave occurring in a rope of length L for which one end is shaken by a hand and the other tied to a door knob. The simplest pattern is shown in figure 12-2a. In this case we produce a standing wave with a wavelength of half the string's length. We call the frequency associated with this wave the fundamental frequency or first harmonic. If the frequency of the hand's oscillations is increased, you can get the second pattern shown in figure 12-2b. In this case the wavelength of the wave is equal to the length of the string. That is the second harmonic. Further increase of the frequency of the hand's oscillations will reproduce the third pattern shown in figure 12-2c or third harmonic. We can generalize this into the formula

$$\lambda_n = \frac{2}{n} L, \tag{4}$$

where $n = 1,2,3,4,5 \ldots$ etc., or since $f = v / \lambda$,

$$f_n = \frac{v}{\lambda_n} = \frac{nv}{2L} = n f_1, \quad \text{where we substituted } \frac{v}{2L} = f_1 \tag{5}$$

Thus successive harmonics f_n are all integral multiples of f_1 the fundamental frequency.

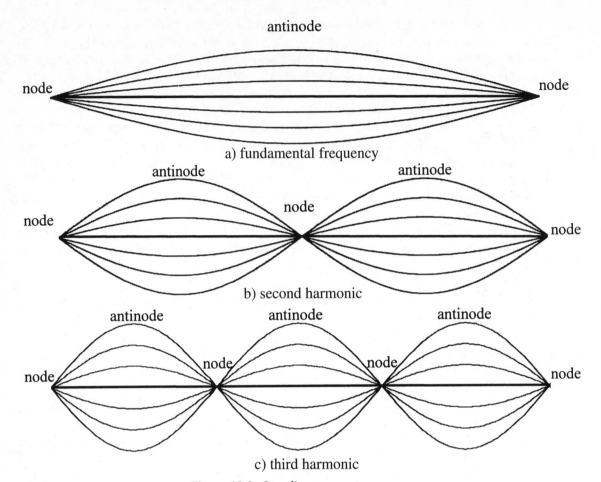

a) fundamental frequency

b) second harmonic

c) third harmonic

Figure 12-2. Standing waves on a rope.

PROCEDURE

PART 1 ■ PULSE

1. Take the slinky and tie a small piece of ribbon to one of its middle coils. Then place the slinky on the floor and have one of your partners hold on to one end of it while you hold the other. Stretch the slinky so that there is some tension in it. Your partner should hold his or her end steady.

2. Have your other partner measure the length of the stretched slinky L and draw chalk marks at its ends perpendicular to the slinky, and a line along the slinky under its middle. Record this length.

$$L =$$

3. By suddenly twitching your hand sideways while holding the slinky create a single pulse in it. You can repeat creating pulses as many times as needed to make observations and measurements. Try to make the pulses as similar as you can by twitching your hand the same speed and distance.

4. With a stop watch measure the time it takes the pulse to reach the other end of the slinky. Calculate the speed of the pulse.

$$t =$$

$$v =$$

5. Have your partner measure the amplitude of the pulse. Record this value.

$$A =$$

6. Observe the motion of the pulse. Observe the motion of the ribbon located at the center of the slinky. Discuss the motion of the pulse and the motion of the ribbon. Does the pulse move along the slinky? Does the piece of the slinky with the ribbon tied to it move along the slinky?

7. Was the disturbance you created a transverse or longitudinal wave pulse?

8. What happens to the pulse when it reaches the opposite end of the slinky?

9. Observe the reflected pulse. What happens to its amplitude upon reflection? Is the reflected pulse inverted? Discuss.

10. Measure the speed of the reflected pulse based on the time it takes the disturbance to travel the length of the slinky and compare it with the speed of the incident pulse (these two measurements should be performed for the same pulse).

Incident pulse   Reflected pulse

$t =$ $t =$

$v =$ $v =$

11. Tie several yards of string to the opposite end of the slinky. Have your partner hold the end of the string. Stretch the slinky to the same length L you had before. Create a pulse in a slinky and observe it. What happens to the pulse when it reaches the opposite end of the slinky? What happens to its amplitude upon reflection? Is the reflected pulse inverted? What happens to the transmitted pulse? Is the transmitted pulse inverted? Discuss.

12. Have your partner untie the string and return to his/her original position. Collect in your one hand a number of turns of the slinky (the more the better) while you hold to the end of the slinky with your other hand. Suddenly release the compressed coils. (You can also try to create a disturbance by rapidly twitching your hand forward or backwards along the slinky while holding its end, but the above procedure usually creates a more pronounced effect) What happens to the disturbance? Measure the speed of the disturbance. Can you identify its amplitude? Discuss.

$$v =$$

$$A =$$

13. Observe the motion of the disturbance. Observe the motion of the ribbon. Discuss the motion of the pulse and the motion of the ribbon. Does the pulse move along the slinky? Does the piece of the slinky with the ribbon tied to it move along the slinky?

14. Was the disturbance you created a transverse or longitudinal wave pulse?

PART 2 ■ SUPERPOSITION

1. Begin with the slinky as it was positioned at the beginning of **PART 1.** The length of the slinky should again be L and its ends matching the chalk marks perpendicular to the slinky and its length along the drawn line.

2. Simultaneously, you and your partner at the other end of the slinky create two pulses both to the same side. Try to make their amplitudes distinctly different so you can distinguish them easily. Observe and discuss what happens before, during, and after the pulses meet. Is it constructive or destructive interference?

3. Simultaneously, you and your partner at the other end of the slinky create two pulses to the opposite side of one another. Try to make their amplitudes the same. Observe and discuss what happens before, during, and after the pulses meet. Is it constructive or destructive interference?

PART 3 ■ STANDING WAVES

1. Begin with the slinky as it was positioned at the beginning of **PART 1.** The length of the slinky should again be L and its ends matching the chalk marks perpendicular to the slinky and its length along the drawn line. Your partner should hold his/hers end steady.

2. By moving your hand the same distance left and right at a uniform pace create a continuous wave on the slinky.

3. Adjust the frequency of your hand motion until you get the standing wave of the fundamental frequency. Have your partner measure the time it takes your hand to complete 10 full oscillations. Record this value. Calculate the period of oscillations.

$$t_{\text{10 oscillations}} =$$

$$T =$$

4. What is the wavelength of the standing wave you created? Record this value.

$$\lambda =$$

5. Sketch the wave you created and indicate the location of nodes and antinodes, the wave's amplitude and wavelength.

6. By moving your hand the same distance left and right at a uniform pace create a continuous wave on the slinky again.

7. Adjust the frequency of your hand motion until you get the standing wave of the second harmonic frequency. Have your partner measure the time it takes your hand to complete 10 full oscillations. Record this value. Calculate the period of oscillations.

$$t_{10 \text{ oscillations}} =$$

$$T =$$

8. What is the wavelength of the standing wave you created? Record this value.

$$\lambda =$$

9. Sketch the wave you created and indicate the location of nodes and antinodes, the wave's amplitude and wavelength.

10. Repeat procedure **PART 3.6–9** for third, fourth and if possible, fifth harmonic. Record your results in table 1 below.

11. In the fourth column of the table calculate the frequency of the waves using formula (1). Then in the fifth column of the table calculate the velocity of the interfering waves. Average this value.

TABLE 1				
n	T_n (s)	λ_n (m)	f_n (Hz)	*v* (m/s)
1				
2				
3				
4				
5				
The average *v* is				

12. With a spring scale measure the magnitude of force **F** required to stretch the slinky to length *L*. Record this value.

$$F =$$

With a balance or electronic scale measure the mass of the slinky. Record this value.

$$m =$$

13. Calculate the velocity of the wave in a slinky using formula (2)

$$v =$$

14. Compare the two values for the velocity by calculating % difference. Within the accuracy of your data can you say that you confirmed relation (2)? Discuss your results.

15. Looking at your data can you say that within experimental accuracy you confirmed relations (4) and (5)?

Electric Circuits Part 2, RC Decay

Purpose

- To investigate the RC circuit discharge.

Equipment and Supplies

- ❏ DC power supply
- ❏ Cables with alligator clips
- ❏ Resistors: 1MΩ
- ❏ Capacitors $C_1 = 500\mu F$, $C_2 = 250\mu F$
- ❏ Two way switch
- ❏ Stopwatch
- ❏ Multimeter

INTRODUCTION

1. Capacitor

A capacitor is a device that stores separated positive and negative electric charges and, thus, energy. The simplest capacitor consists of two parallel conducting plates separated by some distance. One of these plates can then be positively and the other negatively charged. It is found that the amount of charge that can be stored on either the positive or negative plate increases in proportion to the voltage across the plates. Therefore, the capacitance C of the capacitor is defined as the amount of electric charge Q stored on either plate divided by the voltage difference V between the plates

$$C = Q / V, \tag{1}$$

where Q is the charge measured in Coulombs (C), V is the voltage drop measured in Volts (V), and C is the capacitance measured in Farads (F). A Farad is a very large unit and therefore some capacitors have capacitances measured in microfarads denoted by a symbol μF. One μF is equal to one millionth or 10^{-6} of a Farad.

2. RC Circuit

Let us consider a circuit consisting of one resistor and one capacitor connected in series to a power supply as shown in figure 14-1. Such circuits are commonly used as timing devices for example to control the speed of your car's windshield wipers, camera flashes, traffic lights etc. and are commonly referred to as RC circuits.

One more element in the circuit shown in figure 14-1 is an on/off switch which allows us to turn on and off the current flowing through the circuit. Let us start with the switch open and the capacitor uncharged. Once the switch is turned on the current starts to flow through the circuit. The negative charges are transferred from the left plate of the capacitor C through the resistor R and the battery to the right plate of the capacitor C. This will continue until the capacitor is fully charged at which point the current will cease to flow in the circuit. However, it takes a capacitor a very long time (infinitely long strictly speaking) to get fully charged. We define the time constant τ of such circuit as the amount of time it takes the capacitor to gain 63% of its infinite time charge. If the time constant of the circuit is large then the capacitor charges up slowly, and conversely if the time constant of the circuit is small then the capacitor charges up rapidly. The time constant can be calculated by multiplying the capacitance by the resistance i.e.

$$\tau = RC. \tag{1}$$

In equation (1) the resistance should be measured in Ohms (Ω) and the capacitance in Farads (F) to yield the time constant in seconds.

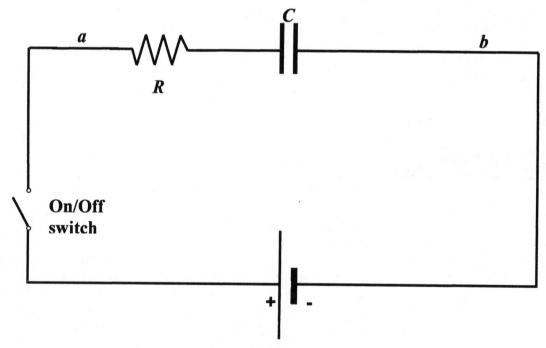

Figure 14-1. RC circuit.

Now that we have a charged capacitor let us remove the battery from the circuit shown in figure 14-1 and reconnect the circuit as in figure 14-2. Once we turn the on/off switch into the on position and close the circuit the capacitor will start to discharge. During discharge the current will start to flow through the circuit and the charge stored on the capacitor will decrease in a manner similar to that in which hot water cools off, i.e. rapidly at first and gradually at slower and slower rate.

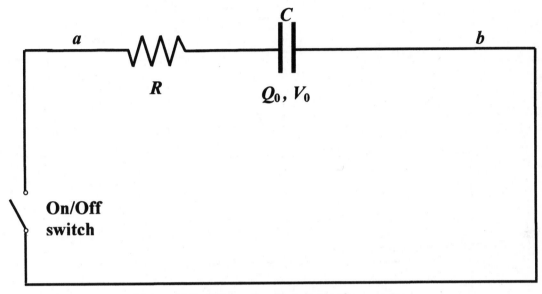

Figure 14-2. RC circuit—discharging the capacitor.

It can be shown that it takes a time

$$t_{1/2} = 0.69 \, RC = 0.69\tau \tag{2}$$

for the voltage to decline to one half its original value.

PROCEDURE

1. Take a capacitor with nominal capacitance of 500 μF and a resistor with nominal resistance of 1 MΩ.

2. Build the circuit as shown in figure 14-3. This setup allows you to both charge and discharge the capacitor i.e. when the switch is in position 1 this circuit is equivalent to the one shown in figure 14-1, so that capacitor can be charged up, and when the switch is in position 2 this circuit is equivalent to the one shown in figure 14-2, so that capacitor can be discharged. At the same time this setup allows you to measure the voltage across the capacitor.

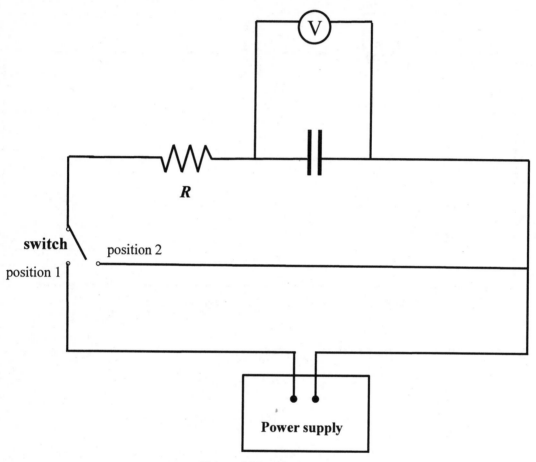

Figure 14-3. RC circuit.

3. With the switch in position 1 turn on the power supply, set it to 10 V, and wait until the voltage across the capacitor stabilizes or does not increase more than 5 mV per minute. That is your original voltage V_0. Record this value in table 1 for the time equal to zero seconds.

4. Simultaneously flip the switch into position 2 and start the stopwatch.

5. Measure and record (in table 1) the voltage every minute for 18 minutes. Turn off the power supply when you finished taking measurements.

TABLE 1

TIME (MINUTES)	TIME (s)	MEASURED VOLTAGE (V)
0	0	
1	60	
2	120	
3	180	
4	240	
5	300	
6	360	
7	420	
8	480	
9	540	
10	600	
11	660	
12	720	
13	780	
14	840	
15	900	
16	960	
17	1020	
18	1080	

6. Using nominal values for R and C_1 calculate the time constant τ of your circuit in seconds.

$$\tau =$$

7. Plot your data on a computer or in the space provided in figure 14-4 and connect the data points with a smooth curve.

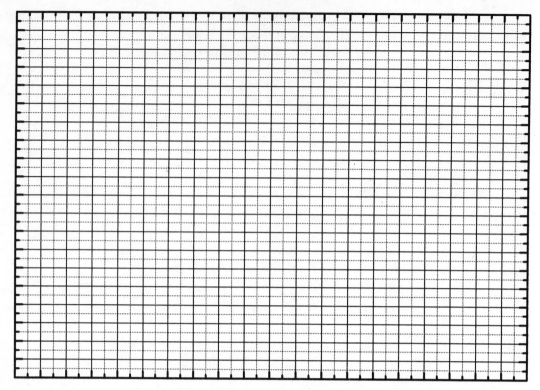

Figure 14-4. Voltage versus time.

8. From your plot deduce how much time it took the voltage across the capacitor to decline from V_0 to $V_0/2$ (i.e. $t_{1/2}$).

$$t_{1/2} =$$

9. From your plot deduce how much time it took the voltage across the capacitor to decline from $V_0/2$ to $V_0/4$ (i.e. $t_{1/2}$).

$$t_{1/2} =$$

10. From your plot deduce how much time it takes took the voltage across the capacitor to decline from $V_0/4$ to $V_0/8$ (i.e. $t_{1/2}$).

$$t_{1/2} =$$

11. Average the three values of $t_{1/2}$ you have just deduced in 8, 9 and 10.

$$t_{1/2} =$$

12. Use equation (3) and the averaged value of $t_{1/2}$ to find your experimental value of the time constant τ. Compare this value of the time constant with the value calculated in step 6 by computing the percent difference. Discuss your results.

$$\tau =$$

13. Replace the 500 μF capacitor with a 250 μF capacitor.

14. With the switch in position 1 turn on the power supply, set it to 10 V, and wait until the voltage across the capacitor stabilizes or does not increase more than 5 mV per minute. That is your original voltage V_0. Record this value in table 2 for the time equal to zero seconds.

15. Simultaneously flip the switch into position 2 and start the stopwatch.

16. Measure and record (in table 2) the voltage every minute until it drops to or below the lowest value you recorded in table 1. Turn off the power supply when you finished taking measurements.

TABLE 2

TIME (MINUTES)	TIME (s)	MEASURED VOLTAGE (V)

17. Using nominal values for R and C_2 calculate the time constant τ of your circuit.

$$\tau =$$

18. Plot your data on a computer or in the space provided in figure 14-4 and connect the data points with a smooth curve. Is the voltage diminishing faster or slower now then it did in the previous circuit?

19. From your plot deduce how much time it took the voltage across the capacitor to decline from V_0 to $V_0/2$ (i.e. $t_{1/2}$).

$$t_{1/2} =$$

20. From your plot deduce how much time it took the voltage across the capacitor to decline from $V_0/2$ to $V_0/4$ (i.e. $t_{1/2}$).

$$t_{1/2} =$$

21. From your plot deduce how much time it took the voltage across the capacitor to decline from $V_0/4$ to $V_0/8$ (i.e. $t_{1/2}$).

$$t_{1/2} =$$

22. Average the three values of $.t_{1/2}$ you have just deduced.

$$t_{1/2} =$$

23. Use equation (3) and the averaged value of $t_{1/2}$ to find your experimental value of the time constant τ. Compare this value of the time constant with the value calculated in step 17 by computing the percent difference. Discuss your results.

$$\tau =$$

24. How does the time constant of the circuit with the 500 μF capacitor compare with the time constant of the circuit with the 200 μF capacitor?

25. How would your results change if you used the 500 μF capacitor and 0.5 MΩ resistor instead?

Induction and Transformers

■ Purpose

- To investigate magnetic induction and the workings of transformers.

☑ Equipment and Supplies

- ❑ 1 permanent bar magnet long enough to be inserted fully into two coaxially placed solenoids.
- ❑ Assortment of solenoids with varying number of turns (recommended $N_1 = 200$ or more, $N_2 = 400$ or more, $N_3 = 800$ or more, $N_4 = 1600$ or more, $N_5 = 3200$ or more)
- ❑ Iron core
- ❑ AC power supply
- ❑ Two multimeters (voltmeters)
- ❑ Cables
- ❑ Compass

INTRODUCTION

Induction is a phenomenon in which a changing magnetic field produces voltage in a wire thus inducing the flow of current.

The transformer is a device which makes it possible to change a small voltage into a large voltage or a large voltage into a small one. They are commonly used as voltage adjusters in power stations and substations, on utility poles and in many electric devices.

A transformer consists of two coils wrapped around an iron core. When a current flows through one of the coils, it magnetizes the iron core. If the current in one of the coils changes the varying magnetic field in the core induces currents in the other coil. We call the coil connected to the AC voltage source the primary. The current in the primary produces an alternating magnetic field in the iron core, which then induces a current in the other coil called the secondary. If the primary coil has $N_{primary}$ turns and the secondary coil has $N_{secondary}$ turns, and we assume that no flux leaks from the iron core (i.e. 100% efficiency), then the voltage induced in the secondary is given by the formula

$$V_{secondary} = (N_{secondary} / N_{primary}) \, V_{primary} \tag{1}$$

Depending on the values of $N_{primary}$ and $N_{secondary}$ we can have a step-up or a step-down transformer i.e. a transformer which either increases or decreases voltage. Generally when the voltage is increased in a coil the current will be decreased. The transformers that you will construct today will generally produce a smaller $V_{secondary}$ than predicted by formula (1) because not all of the magnetic field produced by the primary coil gets transferred by the iron core to the secondary coil. Thus our transformer is not 100% efficient. We can write

$$V_{secondary} = \frac{e}{100\%} \, (N_{secondary} / N_{primary}) \, V_{primary} \tag{2}$$

where e is the efficiency of the transformer in measured in percent.

PROCEDURE

PART 1 ■ INDUCTION

1. If the poles on the bar magnet are unmarked use a compass to identify its north and south poles. Using scotch tape attach small pieces of paper with "S" and "N" written on them to each of the bar magnets so that the poles can be easily identified from now on.

2. Physically connect the solenoid with the largest number of loops that you have available (N_5) and the solenoid with the intermediate number of turns (N_3) to the voltmeters (see figure 16-1). Set the voltmeters to DC current and connect one across each coil.

3. Place the two solenoids side by side so they are coaxial and keep them right next to one another. Make sure their windings are aligned in the same direction. Rapidly insert the north pole of the magnet (see figure 16-1) all the way into the solenoids, pause, and then rapidly remove it (try inserting and removing the magnet at the same speed each time). It is important that the magnet always enters both coils fully. After a few trials you should be able to complete this sequence smoothly. Repeat the action as many times as needed to answer the questions below.

4. Watch the readings of the two voltmeters as you perform the sequence. Based on these readings of the voltmeters is the induced voltage proportional to the number of turns of the solenoid? Discuss.

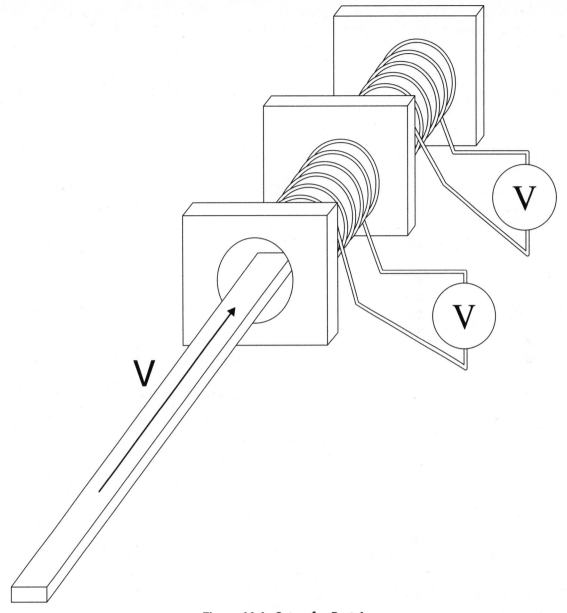

Figure 16-1. Setup for Part 1.

5. Is the sign of the voltage induced when removing the magnet the opposite of that in-duced when inserting the magnet? Discuss.

6. Is there any voltage induced when the magnet is stationary? Discuss.

7. Repeat the action (procedure **PART 1.2**) but insert and remove the magnet at much faster and much slower speeds.

8. Is the voltage induced proportional to the speed of motion? Discuss.

9. Moving the coils rather than the magnet hold the magnet in a horizontal position and rapidly slide the coils so that the magnet passes through them, pause, and then slide them off. It is important that the magnet always enters both coils fully. After a few trials you should be able to complete this sequence smoothly. Repeat the action as many times as needed to answer the question below.

10. Does it make any difference whether it is the magnet or the solenoids that are moved? Discuss.

11. Flip the magnet over and repeat the action (procedure **PART 1.2**) with the south pole inserted first.

12. Does inserting the south pole induce the opposite voltage then inserting the north pole? Discuss.

PART 2 ■ TRANSFORMERS

1. Make a step up transformer with the two solenoids that have the largest number of turns N_4 and N_5 connected on the iron core. Having a step up transformer means that N_4 is your primary and N_5 is your secondary since N_5 is bigger than N_4.

2. Connect the circuit as shown in figure 16-2. Set the voltmeters to AC current and to measure peak-to-peak voltage.

3. Turn the power supply on.

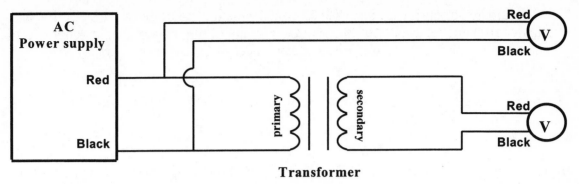

Figure 16-2. Transformer circuit.

4. Set the power supply to a sine wave with amplitude of about 5.0 V peak-to-peak, a frequency of about 60Hz, and DC offset to zero.

5. Measure peak-to-peak voltages for both signals. Record your results in table 1 below.

TABLE 1				
$N_{primary}$	$N_{secondary}$	$V_{primary}$	$V_{secondary}$	e (%)
Average value of e is				

6. Repeat procedure from **PART 2.1** on for step down transformers with $N_{primary} = N_4$ and $N_{secondary} = N_3$, N_2, and N_1. Record your results in table 1 below.

7. Calculate the efficiency for each set of measurements made above and find the average value of e using equation (2). What percentage of the magnetic field generated by the primary coil is transferred to the secondary coil? Discuss your results.

8. If instead of an AC power supply in the experimental setup shown in figure 16-2 you used a DC power supply how would your experiment be affected? Discuss.

| EXPERIMENT | SEVENTEEN |

Reflection, Refraction, and Polarization of Light

◼ Purpose

- To investigate reflection, refraction, and polarization of light.

☑ Equipment and Supplies

- ❑ Flat mirror
- ❑ Cylindrical lens
- ❑ Light source flashlight in a shoebox with a slit cut out
- ❑ Slit plate
- ❑ Slit mask
- ❑ Protractor
- ❑ Ruler
- ❑ Inclined plane
- ❑ Two sheets of polarizer or two pairs of polarizing glasses

INTRODUCTION

1. Reflection

All objects reflect some of the light incident upon them. It is this reflected light which allows us to see the objects. Many surfaces are very irregular on a microscopic scale and reflect light almost equally in all directions producing what is called diffuse reflection. Surfaces sufficiently smooth so that any irregularities are small compared to the wavelength of light reflect light in basically one direction. This is called specular reflection which occurs, for example, when you observe an image in a mirror.

When light is reflected specularly, the angle of incidence θ_i equals the angle of reflection θ_r (see figure 17-1), where both of these angles are measured with respect to the line perpendicular to the mirror's surface called the normal.

$$\theta_i = \theta_r \qquad \textit{Law of reflection} \qquad (1)$$

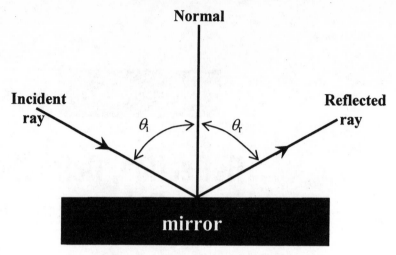

Figure 17-1. Reflection of a ray.

2. Refraction

When light passes from one medium into another, its path is refracted or bent. That is why a pencil inserted in a water container appears bent. This phenomenon results from the fact that light travels at different speeds in different media. Let us define the index of refraction n of a given medium as the ratio between the speed of light in free space, c, and the speed of light in this particular medium, v, i.e.

$$n = \frac{c}{v}. \tag{2}$$

Snell's law relates the angle of incidence θ_1 and the angle of refraction θ_2 (i.e. the angles the incident and refracted light makes with the normal (see figure 17-2)) to the ratio of the refractive indexes of the two media n_1 and n_2 in the following way

$$\frac{\sin\theta_1}{\sin\theta_2} = \frac{n_2}{n_1} \qquad \textbf{\textit{Law of refraction (Snell's Law)}} \tag{3}$$

(if you are unfamiliar with trigonometry and do not know what sin (or sine) of an angle is read the appendix for explanation)

From Snell's Law, it follows that if light travels from some medium with a relatively high index of refraction into a medium of lower index of refraction, the light is bent away from the normal and the refracted angle is greater than the incident angle. At some angle, called the critical angle (θ_c) the refracted ray will emerge along the interface between the media (see ray B in figure 17-3). At any angle of incidence greater than θ_c, the light will be totally internally reflected.

We can find the critical angle θ_c, by noting that the refracted ray for this angle of incidence is at 90° to the normal. Thus the Snell's law becomes

$$\frac{\sin\theta_c}{\sin 90°} = \frac{n_1}{n_2} \tag{5}$$

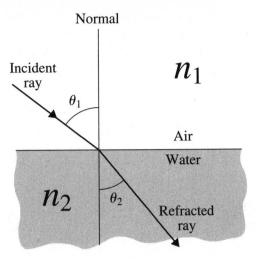

Figure 17-2. Refraction of a ray.

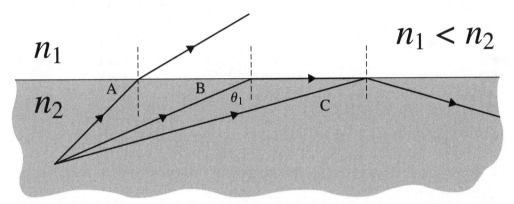

Figure 17-3. Critical angle: ⁓ ⁓ **strikes the interface at an angle less than critical angle; ray B strikes the interface** ⁓ ⁓ ⁓ ⁓ ⁓ **critical angle; ray C strikes at greater than critical angle and is therefore totally internally reflected.**

since the incident ray now travels through the medium with higher refractive index n_2. The sin 90° is equal to1, so we obtain

$$\sin \theta_c = \frac{n_1}{n_2} \tag{6}$$

If the medium of lower index of refraction is air (which index of refraction is 1) then

$$\sin \theta_c = \frac{1}{n_2}. \tag{7}$$

3. Polarization

To illustrate the phenomenon of polarization let us first consider a wave traveling in a rope. Let us assume that the oscillations in the rope can be induced only in a vertical or only in a horizontal plane (see figure 17-4a and 17-4b). If you now place a vertical slit in the path of such waves only vertically polarized waves would pass through (see figure

17-4c and 17-4d). Similarly, if you place a horizontal slit in the path of these wave only horizontally polarized waves would pass through. If both slits are used, both types of waves will be stopped. Notice that only a transverse and not longitudinal waves can be polarized, thus our ability to polarize light proves that, unlike sound waves, light is a transverse wave.

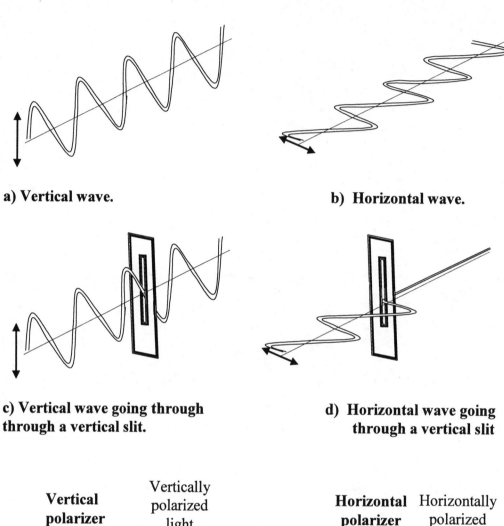

a) Vertical wave.   **b) Horizontal wave.**

c) Vertical wave going through through a vertical slit.   **d) Horizontal wave going through a vertical slit**

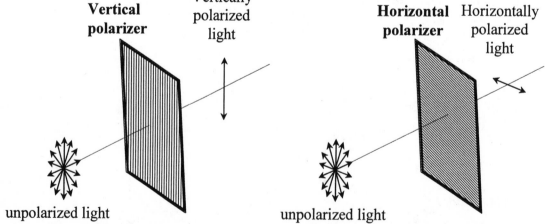

e) Polarization and electric fields.   **f) Polarization and electric fields.**

Figure 17-4. Polarization

Light being a transverse wave means that the electromagnetic disturbances that compose light occur in a direction perpendicular to the direction of propagation. Specifically, polarization for light, refers to the orientation of the electric field in the electromagnetic wave. The magnetic field in an electromagnetic wave is always perpendicular to the electric field. Figure 17-4e and 4f show vertical and horizontal polarization, respectively.

Most light sources such as sunlight or an incandescent bulb produce light that is unpolarized. That is because a beam of light produced by such source consists of a multitude of waves emitted by molecules vibrating in all possible directions. Polarization of light can be achieved by placing a special material called a polarizer in the beams path. A polarizer filters out all but one preferred orientation of oscillation by acting as a series of parallel very narrow slits. Polarization can also be achieved by reflection. For example when sunlight strikes a surface of a lake at an angle other then perpendicular, the reflected beam becomes polarized to some extent in the direction parallel to the surface. That is why people use polaroid glasses in order to avoid glare.

PROCEDURE

PART 1 ■ REFLECTION

1. On a sheet of paper draw two lines perpendicular to each other. Label one of them "mirror" and the other "normal".

2. Set up the lab equipment provided as shown in figure 17-5.

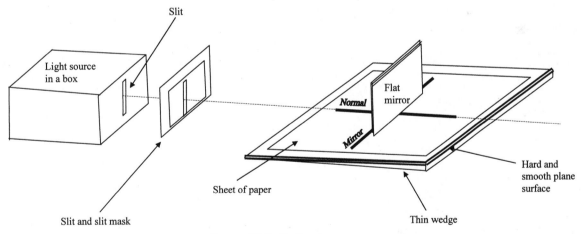

Figure 17-5. Reflection setup.

3. Adjust the components so a single ray of light is aligned with the normal.

4. Carefully align the reflecting surface of the mirror with the line labeled mirror. The incident and reflected rays should both follow the normal line.

5. Gently rotate the paper without sliding it sideways and observe the light rays. Do not move the inclined plane itself. The angles of incidence and reflection are measured with respect to the normal to the reflecting surface, as shown in figure 17-1.

6. For 3 different incidence angles measure and record the angles of incidence and reflection. You can do this by tracing the rays on your sheet of paper and then sliding the sheet out and using the protractor to measure the angles. Make sure that the incident ray hits the mirror at the intersection of the mirror line and the normal line. Record the measured values for the angles of incidence and reflection in table 1 below.

TABLE 1

θ_i (°)	θ_r (°)

7. Are the angles of incidence and reflection the same considering accuracy of your measurement? Did you verify the law of reflection given by equation (1)? Discuss.

PART 2 ■ **REFRACTION**

1. On a sheet of paper draw two lines perpendicular to each other. Label one of them "lens" and the other "normal".

2. Set up the lab equipment provided as shown in figure 17-6.

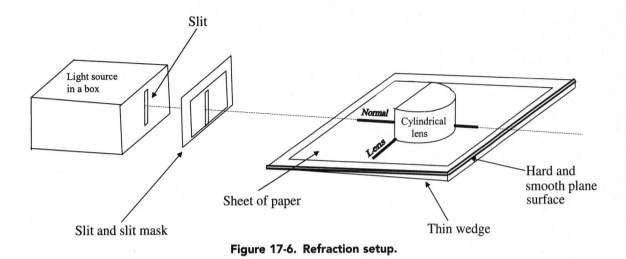

Figure 17-6. Refraction setup.

3. Adjust the components so a single ray of light is aligned with the normal line.

4. Carefully align the flat surface of the cylindrical lens with the line labeled lens. The incident and refracted rays should both follow the normal line and strike the flat surface of the lens in its middle.

5. Gently rotate the paper without sliding it sideways or disturbing the alignment of the lens and observe the light rays. Do not move the inclined plane itself. The angles of incidence and refraction are measured with respect to the normal line to the lens' flat surface, as shown in figure 17-7.

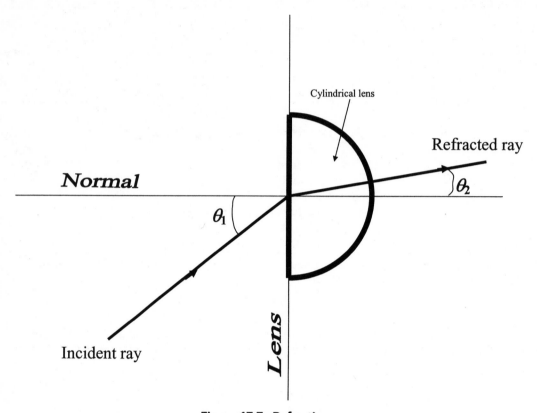

Figure 17-7. Refraction.

6. Is the ray bent when it passes into the lens perpendicular to the flat surface of the lens? Is the ray bent when it passes out of the lens perpendicular to the curved surface of the lens? If the ray is bent on the curved surface then your lens is off center. Make sure that the incident ray hits the middle of the flat surface.

7. What medium is the incident ray traveling in?

8. What medium is the refracted ray traveling in?

9. Is the refracted ray bent away or towards the normal? Explain why.

10. By tracing the rays on the paper as you did in **PART 1** measure and record (in table 2) the angles of incidence and refraction for 3 different incidence angles lying on the same side of the normal.

TABLE 2

θ_1 (°)	θ_2 (°)

11. Repeat the measurements with the incident ray striking from the opposite side of the normal for the same 3 angles of incidence. Record your results in table 3. Are your results for the two sets of measurements the same? If not, to what do you attribute the differences?

TABLE 3

θ_1 (°)	θ_2 (°)

12. In table 4 for each of the three angles, use the law of refraction (equation (3)) to calculate the index of refraction of the glass that the cylindrical lens is made out of. Assume that the index of refraction for air is equal to 1. Are the values consistent? Calculate the average value of the index of refraction n. Discuss your results.

TABLE 4

MEASUREMENT	n
1	
2	
3	
The average value of n is	

13. Use the average value of n and equation (2) to find the speed of light in the glass that the cylindrical lens is made out of. Assume that $c = 3 \times 10^8$ m/s

$$v =$$

PART 3 ■ TOTAL INTERNAL REFLECTION

1. Set up the equipment as you did in **PART 2** but place the flat surface of the cylindrical lens facing away from the light source as shown in figure 17-8.

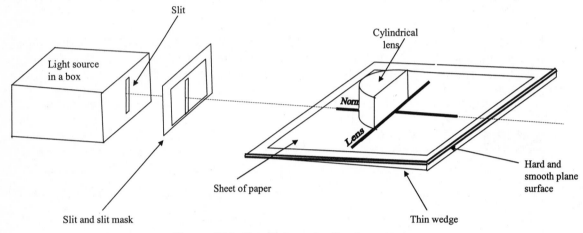

Figure 17-8. Total internal reflection setup.

2. Gently rotate the paper without sliding it sideways or disturbing the alignment of the lens and observe the light rays.

3. From which surface of the lens front or back does reflection now primarily occur? If the ray is bent on the curved surface then your lens is off center. Make sure that the incident ray hits the middle of the flat surface.

4. What medium is the incident ray traveling in?

5. What medium is the refracted ray traveling in?

6. Is the refracted ray bent away or towards the normal? Explain why.

7. Is there a reflected ray for all angles of incidence?

8. Is there a refracted ray for all angles of incidence?

9. How do the intensities of the reflected and refracted rays vary with the angle of incidence?

10. At what angle is all the light reflected (i.e. total internal reflection occurs)? Use equation (7) to determine the index of refraction of the glass that the lens is made out of. How does this value compare with the value of the index of refraction calculated in **PART 2**?

$$n =$$

PART 4 ■ POLARIZATION

Your optics equipment includes two polarizers, which transmit only light that is polarized along a specific axis. Light that is polarized along any other direction is absorbed by the polaroid material. Therefore, if unpolarized light enters the polarizer, the light that passes through is polarized along the polarization axis.

1. Turn the light source on and view it without polarizers.

2. Place polarizer 1 in the path of light. Rotate the polarizer while viewing the light source through it. Note and discuss the changes in light's intensity. Is the light from the light source polarized? How can you tell?

3. Place polarizer 2 in the path of light. Hold polarizer 1 steady and looking through both polarizers, rotate polarizer 2. Are there some angles of polarizer 2 for which maximum of light is transmitted? Are there some angles of polarizer 2 giving no light transmitted? Explain.

Appendix

Trigonometry is a part of mathematics based on the properties of right triangles i.e. triangles which two sides are at 90° to each other. A sample right triangle is shown in figure 17-A1 below. In this triangle side a is opposite the angle θ, side b is adjacent to the angle θ, and the side c, opposite the 90° angle, is the hypotenuse of the triangle. The ratio's of the lengths of the sides of such triangle define basic trigonometric functions such as sine (sin), cosine (cos), tangent (tan) of the angle θ in the following way:

$$\sin \theta = \frac{\text{length of the side opposite } \theta}{\text{length of the hypotenuse}} = \frac{a}{c}$$

$$\cos \theta = \frac{\text{length of the side adjacent to } \theta}{\text{length of the hypotenuse}} = \frac{b}{c}$$

$$\tan \theta = \frac{\text{length of the side opposite to } \theta}{\text{length of the side adjacent to } \theta} = \frac{a}{b}$$

Since sine, cosine, and tangent are ratios of two lengths they are quantities that do not have any units.

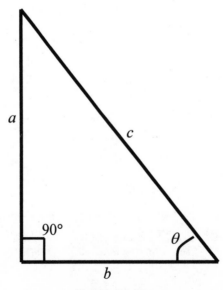

Figure 17-A1.

The values of sine, cosine, and tangent for any particular angle θ can now be easily found using calculator. Make sure you set your calculator in a proper mode to enter the angle in degrees. In table A1 below we list values of sine, cosine, and tangent for few frequently used angles. See if you get these values with your calculator.

TABLE A1

ANGLE θ (°)	$\sin \theta$	$\cos \theta$	$\tan \theta$
0	0	1	0
30	$\dfrac{1}{2}$	$\dfrac{\sqrt{3}}{2} \approx 0.866$	$\dfrac{1}{\sqrt{3}} \approx 0.577$
45	$\dfrac{\sqrt{2}}{2} \approx 0.707$	$\dfrac{\sqrt{2}}{2} \approx 0.707$	1
60	$\dfrac{\sqrt{3}}{2} \approx 0.866$	$\dfrac{1}{2}$	$\sqrt{3} \approx 1.732$
90	1	0	∞

Mirrors and Lenses

▯ Purpose

- To investigate some of the properties of mirrors and lenses.

☑ Equipment and Supplies

- ❏ Flat mirror
- ❏ Cylindrical lens
- ❏ Concave spherical mirror
- ❏ Convex spherical mirror
- ❏ Thin converging lens (1) e.g. f = +150 mm
- ❏ Thin diverging lens (1) e.g. f = −150 mm
- ❏ Thin converging lens (2) e.g. f = +50 mm
- ❏ Cross arrow target (opaque cardboard with two perpendicular intersecting arrows of the same size cut out)
- ❏ Viewing screen (rectangular piece of opaque cardboard)
- ❏ Planoconvex cylindrical thin lens
- ❏ Light source flashlight in a shoebox with a slit cut out
- ❏ Slit plate
- ❏ Slit mask
- ❏ Ruler
- ❏ Inclined plane
- ❏ A stand for lenses

INTRODUCTION

1. *Mirrors*

The law of reflection you studied in the previous lab applies also to mirrors. In part of this experiment you will examine reflection from flat, convex and concave spherical mirrors.

2. Lenses

The law of refraction you studied in the previous lab applies also to lenses. In part of this experiment you will further examine the refraction occurring with converging and diverging lenses.

Converging lenses are those which are thicker in the middle then at the edges as shown in figure 18-1. They all focus light at a point on the side of the lens opposite from a distant light source.

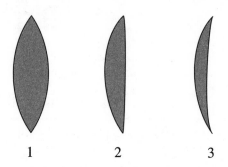

1   2   3

Figure 18-1. Three types of converging lenses.

Diverging lenses are those which are thinner in the middle then at the edges as shown in figure 18-2. They all focus light so that for a viewer on the side of the lens opposite that from which light is incident, the object appears to be at a point behind the lens, i.e., the same side as the light source.

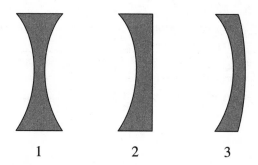

1   2   3

Figure 18-2. Three types of diverging lenses.

With the proper combination of lenses, various optical devices such as eyeglasses, cameras, telescopes, microscopes, etc., may be constructed. In such instruments light is focused to produce the desired image.

For spherical lenses, there is a general equation that can be used to determine the location and magnification of an image. This equation called the lens equation is written:

$$\frac{1}{d_i} + \frac{1}{d_o} = \frac{1}{f},$$

(1)

here f is the focal length of the lens, and d_o and d_i are the distances from the lens to the object and image, respectively (see figure 18-3). The magnification of the image by a lens is defined as

$$M = \frac{h_i}{h_o}. \tag{2}$$

where h_o and h_i are the heights of the object and the image, respectively. It can be shown that the magnification is also equal to the ratio of image and object distances

$$M = \frac{d_i}{d_o}. \tag{3}$$

In this experiment, you will have an opportunity to test these three equations.

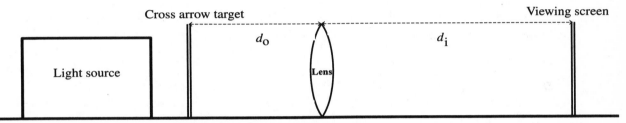

Figure 18-3. Setup for spherical lens experiment.

PROCEDURE

PART 1 ■ RAY TRACING

In this section you will observe the passage of parallel rays of light through several optical devices.

1. Set up the equipment as shown in figure 18-4 with the slit plate and the planoconvex cylindrical thin lens behind it. Adjust their position with respect to the light source so that parallel rays come through and are centered on the normal drawn on the paper sheet. The sheet of paper will serve as a screen on which you view the paths of the rays.

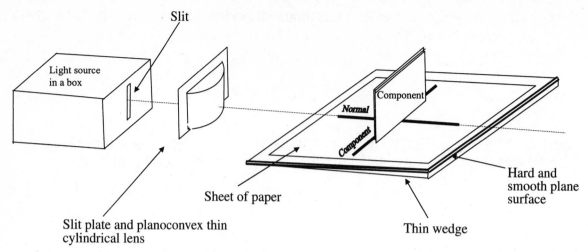

Figure 18-4. Ray tracing setup.

2. Sketch the ray patterns as the light passes through or reflects off the following devices and discuss how each of these devices affects the paths of parallel rays:

- the flat mirror placed at the component line;

- the convex mirror placed at the component line;

- the concave mirror placed at the component line;

- the cylindrical lens placed with the flat surface at the component line;

- the cylindrical lens placed with the convex surface at the component line;

- the converging lens 1 placed between the slit plate and the edge of the paper facing the light. Measure the focal length of this lens by measuring the distance between the edge of the lens and the point where parallel rays converge. Record this value;

$$f =$$

- the converging lens 2 placed between the slit plate and the inclined plane. Measure the focal length of this lens and record it;

$$f =$$

- the diverging lens placed between the slit plate and the inclined plane;

- the converging lens 2 and the diverging lens placed together back to back between the slit plate and the inclined plane.

PART 2 ■ THIN LENSES

1. Setup the equipment as shown in figure 18-3. Use thin converging lens 1.

2. Turn on the light source and slide the lens toward or away from the crossed arrow target as needed to focus the image of the target onto the viewing screen.

3. Is the image magnified or reduced in size?

4. Is the image inverted or upright?

5. What happens to the image distance d_i if you increase the objects distance d_o even further?

6. What happens to the image distance d_i if the objects distance d_o were very, very large?

7. Now, for 5 different object distances d_o between the lens and the crossed arrow target measure the distance between the lens and the image d_i (or viewing screen), as well as the height of the image. At each position, place the viewing screen so the image of the target is in sharp focus. Record your data in table 1.

TABLE 1

d_o (mm)	d_i (mm)	h_i (mm)	f (mm)	$M = \dfrac{h_i}{h_o}$	$M = \dfrac{d_i}{d_o}$
average focal length f of the thin converging lens 1 is					

8. Using equation (1) find the value of the focal length f for each measurement and then find its average.

9. Are your values for the focal length from the five object distances consistent within experimental errors? If not, to what do you attribute the discrepancies?

10. Compare your average value for f with the one measured in **PART 1.** Discuss your results.

11. Did you verify the lens equation?

12. In table 1 calculate the magnification using both the height ratio (equation 2) and distance ratio (equation 3). Are the numbers similar? Which do you consider more accurate and why? How does the magnification change with decreasing object distance d_o?

13. For what values of object distance d_o were you unable to focus an image onto the screen? Use the lens equation to explain why.

Additional Questions (Challenge):

1. According to the lens equation, what value of the object distance d_o would get an image with a magnification of one?

2. Is it possible to obtain a non-inverted image with a converging spherical lens? Explain.

3. For a converging lens of focal length f, where would you place the object to obtain an image as far away from the lens as possible? How large would the image be?

EXPERIMENT **NINETEEN**

Atomic Spectra

▮ Purpose

- To investigate atomic spectra using a diffraction grating.

☑ Equipment and Supplies

- ❏ Diffraction grating (recommended slit spacing of about 5.5×10^{-6} m or about 1800 slits/cm)
- ❏ 2 Flashlights
- ❏ Laser pointer
- ❏ Fluorescent light source
- ❏ Hydrogen gas tube with its power supply
- ❏ Tubes with several other gases (e.g. helium, sodium, mercury, neon, oxygen)
- ❏ Spectral charts for these gases
- ❏ Meter stick
- ❏ Slit and slit mask
- ❏ Wooden blocks

INTRODUCTION

In this lab we will study the light emitted by hydrogen and other gases contained in special glass tubes. When energy, in the form of electricity, is introduced into the gas tube, the electrons in the gas absorb that energy and become excited or in other words they move from their lowest energy levels, so called ground states, to higher energy levels called excited states (see figure 19-1). In order to move to an excited state, an electron needs energy. It will absorb this energy in discrete amounts according to the nature of atomic systems. The electron will then decay (or fall back) from the excited state and return to its ground or an intermediate excited state while emitting a photon of a particular energy in the process. This "de-excitation" results in the emission of light with particular wavelength which can be measured. The process goes on continuously as long as energy is provided to the gas. In this experiment, we will look at light emission from several elemental gases.

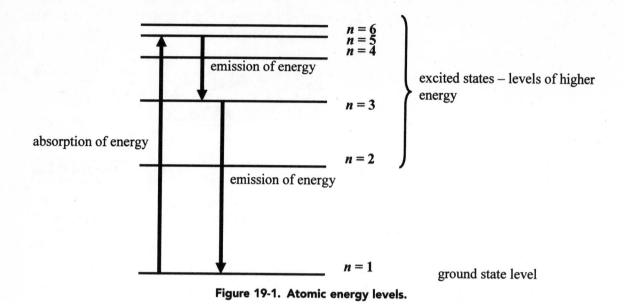

Figure 19-1. Atomic energy levels.

Question

Since each element has a unique combination of electrons in various energy levels, would you think that the characteristic wavelengths or so called emission lines of different elements should be identical or different? Could these elements have some lines which are the same and yet the total spectra not the same?

Since all of the emission lines produced by a light source are combined together, tools must be used to separate the visible light into its spectral lines or distinct light colors. Separation of the spectral lines can be achieved in a number of methods. We see light separated into components after a rainstorm through droplets of moisture in the sky producing a rainbow much like a prism disperses light. When the white light travels through a prism or a droplet of water, the different light wavelengths will be refracted at different angles. In today's lab you will use a diffraction grating to separate the different wavelengths of light into its color spectra. The diffraction grating has thousands of scratches or slits that the light travels through. The light wave that passes through each slit combines with the light from all the other slits. These separate light waves will all be in step with each other only for certain specific angles, and a given wavelength light will produce a line in the spectrum at these angles. The lines are actually images of the slit. The equation for the angle θ of deviation of any wavelength λ is

$$\sin \theta = \frac{m\lambda}{D} \tag{1}$$

where $m = 0,1,2,3$ is known as the order number of the diffraction pattern and D is the slit separation distance of the grating (see figure 19-2). Notice that for the $m = 0$ order all wavelengths combine to give an image of the slit at $\sin \theta = 0$ or at zero degrees.

One of the milestones in the development of our understanding of the atom was the empirical Balmer formula giving the wavelengths of light emitted from dilute hydrogen gas. The formula is written as:

$$\lambda = \frac{1}{R\left(\dfrac{1}{2^2} - \dfrac{1}{n^2}\right)} \text{ with } n = 3, 4, 5, \ldots \text{ (an integer)} \tag{2}$$

where λ is the wavelength of the hydrogen emission line and $R = 1.097 \times 10^7 \text{ m}^{-1}$ is the Rydberg constant.

For elements other than hydrogen there is no corresponding simple formula for the wavelengths in the spectra, but the spectral charts show the measured wavelengths for some common cases.

PROCEDURE

PART 1 ■ DIFFRACTION GRATING

1. To get an idea of what the diffraction grating does pick one up and look through it. Describe and explain what you see.

2. Take a flash light, turn it on. Look through the diffraction grating directly at the light produced by the flashlight. Describe and explain what you see.

3. Take a laser pointer and turn it on. **CAUTION: DO NOT LOOK DIRECTLY INTO THE BEAM OR POINT IT AT SOMEONE ELSES' EYES.** Look through the diffraction grating at the dot produced by the pointer on the wall. Describe and explain what you see. What are the similarities and differences between what you observe with the laser and the flashlight? Explain.

4. Look through the diffraction grating at the fluorescent light fixtures. Describe and explain what you see.

PART 2 ■ HYDROGEN SPECTRUM

CAUTION: Hydrogen and other atomic gas tubes require high voltage. You must not touch the tube holder and the holder should be unplugged when you change the tubes. Also the tubes get hot, so exercise caution.

1. Now that you know what the colored bands of light—the spectrum—should look like, assemble the equipment as shown in figure 19-2. Place the slit mask in front of the flashlight in such way that it produces a single beam of light.

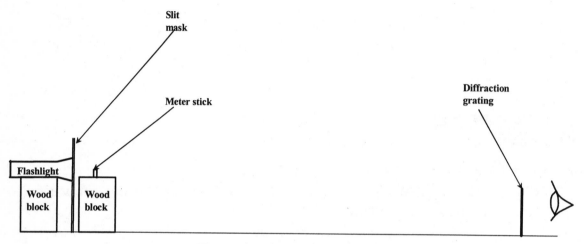

Figure 19-2. Experimental setup.

2. Look through the diffraction grating at the light source and locate (on the meter stick) the $m = 0$, $m = 1$, $m = 2$, etc. spectra on each side. Your partner may need to move a pen along the meter stick to where you see it overlap with the desired part of the spectra. Then you can use the other flashlight to read the position on the meter stick. How many spectra can you see on each side of $\theta = 0$? How does the intensity of the light change with increasing m? Why? Discuss.

3. Select a hydrogen gas tube and place it in its power supply making sure it is in properly. Replace the flashlight with the hydrogen gas tube and remove the slit mask so that your apparatus is assembled as shown in figure 19-3. Turn on the power supply. Place the meter sick on top of the wooden block so that it rests at the height that matches the middle of the hydrogen tube. Make sure the 50 cm mark is aligned with the middle of the tube.

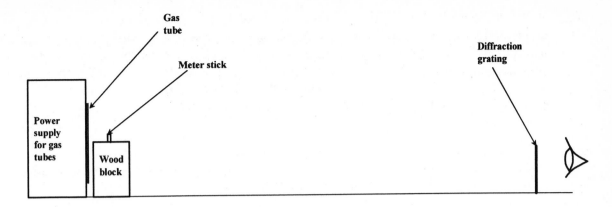

a) Side view

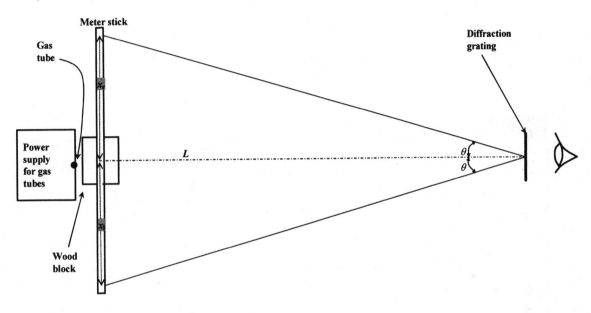

b) Top view

Figure 19-3. Experimental setup.

4. Measure the distance L from the grating to the meter stick. Record this value. Do not change this distance for the remainder of the experiment.

$$L =$$

5. Look through the diffraction grating at the diffraction pattern and locate (on the meter stick) the $m = 0$, $m = 1$, $m = 2$, etc. spectra on each side. How many different orders can you see? How many emission lines can you see in each order? Some people are able to see 4 some only 3 lines. Why? Are any lines missing from your data when compared to the spectral chart? Why might this be?

6. For $m = 1,2$ and 3 measure the position of each spectral line on the meter stick to the left and to the right of the light source. Again you will need your partners help with marking the position along the meter stick. From these positions calculate the distance along the meter sitck between each spectral line and the central 50 cm mark (or $\theta = 0$). For each m we will denote the distance to a particular spectral line to the left as x_l and to the right as x_r. Record your results in the table 1.

7. Average the distance to the right and left of $\theta = 0$ by calculating $x = (x_l + x_r)/2$, and record it in table 1. Then calculate $\sin \theta$ from the relation

$$\sin \theta = \frac{x}{a} = \frac{x}{\sqrt{x^2 + L^2}}.$$

8. Use formula (1) to find the wavelength of each spectral line from its angle θ. Record your results in table 1.

TABLE 1

m	COLOR	x_l (mm)	x_l (mm)	AVERAGE x (mm)	$\sin \theta$	λ (nm)
1	Red					
	Blue					
	Dark Blue					
	Violet					
2	Red					
	Blue					
	Dark Blue					
3	Violet					
	Red					
	Blue					
	Dark Blue					
	Violet					

9. Average your values for the wavelength of each spectral line (you have three values one for each m) and record this value in table 2. Assign each wavelength a quantum number n. (Hint: the longest wavelength line has $n = 3$.)

10. Use the Balmer formula (equation (2)) to predict the theoretical values of the wavelengths of the spectral lines. Compare your experimental and theoretical values by calculating the % difference. Record your results in table 2. Within your experimental accuracy did you confirm the Balmer formula? Discuss your results.

TABLE 2

COLOR	AVERAGE EXPERIMENTAL WAVELENGTH λ (nm)	QUANTUM NUMBER n	THEORETICAL BALMER WAVELENGTH λ (nm)	% DIFFERENCE
Red				
Blue				
Dark Blue				
Violet				

PART 3 ■ **ATOMIC SPECTRA**

Look at the spectra of light emitted by the gas tubes containing different gases. For each element record the colors the most prominent emission lines you can see and compare them to the ones on the spectral chart provided in the lab room. Try to identify their wavelengths. Use table 3 below to organize your data. Are any lines missing from your data when compared to the chart? Why might this be? Discuss your results.

TABLE 3

ELEMENT	LINE COLOR	λ (nm)

Nuclear Decay Simulation

⬛ Purpose

- To enhance the understanding of half-life and radioactive decay using dice to simulate radioactive nuclei.

☑ Equipment and Supplies

- ❑ Dice (200 or more)
- ❑ A box with a lid large enough to allow the dice to move freely inside while shaken.

INTRODUCTION

Many things decrease at a diminishing rate such as the charge on a discharging capacitor, the temperature of a cooling object, the amount of certain nuclear isotopes during radioactive decay.

Radioactive materials are characterized by their rates of decay or their half lives. If initially you start out with N_0 radioactive nuclei in your sample of radioactive material with a half life of $t_{1/2}$ then after the passage of that amount of time approximately half of them will spontaneously decay. It is a statistical phenomenon—we cannot really say that this particular nucleus will decay, we only know that over the time of one half life it is likely that half of them will decay. Of the remaining $N_0/2$ nuclei approximately one half will decay during the subsequent period of time $t = t_{1/2}$, and so forth.

In this experiment we will be using dice to represent atomic nuclei. Dice having certain faces up will represent radioactive nuclei, and dice having other faces up will represent nuclei that have decayed, and are no longer radioactive, i.e., are stable. We will simulate the process of radioactive decay by placing a large number of dice in a container and shaking it—allowing some to make the transition from "radioactive" to "stable." One shake of the container simply represents the passage of a certain amount of time. To make the analogy fit we need to remove dice that have become stable after each shake, because otherwise we would find some dice making the transition from stable back to radioactive, which cannot occur with nuclei. The dice analogy fits radioactive decay very well because whether or not a given nucleus will decay in the next one second does not depend on how long it has been "waiting" to decay; likewise whether a die will flip in the next shake doesn't depend on how many shakes have occurred previously.

PROCEDURE

PART 1 ■

1. Pick a single die and roll it on the table. If the die ends up with "1'" up we say it "decayed". Roll the die several times, until you can answer the following questions:

    a) how long (i.e. how many throws) on average before the die "decays"?

    b) what is the chance of a die decaying on each particular throw?

2. Count all your dice and record that number (N_0).

$$N_0 =$$

3. Place the dice in the box and close it with the lid.

4. Shake the box thoroughly in all directions and turning it upside down.

5. Open the box, remove and count all the dice with "1'"s up. These represent the nuclei that have decayed. Do not put the removed dice back in the box. Record your result in table 1. Replace the lid.

6. Repeat the procedure step 4 and 5 until less than 3 dice remain in the box. Record your result in table 1.

7. In the third column of your table calculate the number of dice remaining in the box after each shake.

TABLE 1

SHAKE NUMBER	NUMBER OF REMOVED DICE AFTER THIS SHAKE	NUMBER OF DICE REMAINING IN THE BOX
0	0	
1		
2		
3		
4		
5		
6		
7		
8		
9		
10		
11		
12		
13		
14		
15		
16		
17		
18		
19		
20		
21		
22		
23		
24		
25		

8. Plot the number of dice remaining in the box for each shake versus the shake's number. You can use your computer or the space provided in figure 20-1. Remember to label the axis. Connect the data points with a curve.

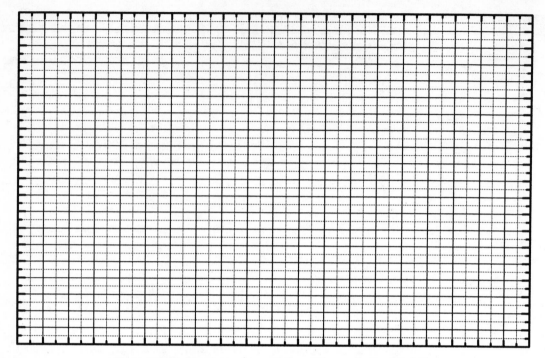

Figure 20-1. Number of remaining dice vs. shake number.

9. What is the meaning of the graph you have just plotted in the context of nuclear decay?

10. Approximately what percent of the dice remaining in the box were removed at each shake? What percent remained? Why?

11. How many times did you shake the box before half of the dice were removed?

12. From your plot deduce how many shakes it took for the number of the dice remaining in the box to decline from N_0 to $N_0/2$ (i.e. $t_{1/2}$)?

$$t_{1/2} =$$

13. From your plot deduce how many shakes it took for the number of the dice remaining in the box to decline from $N_0/2$ to $N_0/4$ (i.e. $t_{1/2}$)?

$$t_{1/2} =$$

14. From your plot deduce how many shakes it took for the number of the dice remaining in the box to decline from $N_0/4$ to $N_0/8$ (i.e. $t_{1/2}$)?

$$t_{1/2} =$$

15. Average the three values of .half-life for the dice you have just deduced.

$$t_{1/2} =$$

PART 2 ■

In this part of the experiment we are going to simulate the nuclear decay process using a different half-life than in **PART 1**.

1. Pick a single die and roll it on the table. If the die ends up with "1" up or with "2" up we say it "decayed". Roll the die several times, until you can answer the following questions:

a) how long (i.e. how many throws) on average before the die "decays"?

b) what is the chance of a die decaying on each particular throw?

2. Place the dice in the box and close it with the lid.

3. Shake the box thoroughly in all directions and turning it upside down.

4. Open the box, remove and count all the dice with "1"'s up and with "2"'s up. These represent the nuclei that have decayed. Do not put the removed dice back in the box. Record your result in table 2.

TABLE 2

SHAKE NUMBER	NUMBER OF REMOVED DICE AFTER THIS SHAKE	NUMBER OF DICE REMAINING IN THE BOX
0	0	
1		
2		
3		
4		
5		
6		
7		
8		
9		
10		
11		
12		
13		
14		
15		
16		
17		
18		
19		
20		
21		
22		
23		
24		
25		

5. Repeat the procedure **PART 2** step 3 and 4 until less than 3 dice remain in the box. Record your result in table 2. Replace the lid.

6. In the third column of your table calculate the number of dice remaining in the box after each shake.

7. Plot the number of dice remaining in the box for each shake versus the shake's number. You can use your computer or the space provided in figure 20-1. Connect the data points with a curve.

8. How does this curve differ from the one you plotted in **PART 1**? Compare the two curves.

9. Approximately what percent of the dice remaining in the box were removed at each shake? What percent remained? Why?

10. How many times did you shake the box before half of the dice were removed?

11. From your plot deduce how many shakes it took for the number of the dice remaining in the box to decline from N_0 to $N_0/2$ (i.e. $t_{1/2}$)?

$$t_{1/2} =$$

12. From your plot deduce how many shakes it took for the number of the dice remaining in the box to decline from $N_0/2$ to $N_0/4$ (i.e. $t_{1/2}$)?

$$t_{1/2} =$$

13. From your plot deduce how many shakes it took for the number of the dice remaining in the box to decline from $N_0/4$ to $N_0/8$ (i.e. $t_{1/2}$)?

$$t_{1/2} =$$

14. Average the three values of .half-life for the dice you have just deduced.

$$t_{1/2} =$$

15. Compare the half lives for you obtained in **PART 1** and in **PART 2**. Is the half life decreasing or increasing as the probability of decay increases? Discuss your results.

Notes

Notes

Notes

Notes

Notes

Notes